Handbook of Power Electronics

Handbook of Power Electronics

Contributors

M. S. Ali, S. K. Kamarudin et al.

www.aurisreference.com

Handbook of Power Electronics

Contributors: M. S. Ali, S. K. Kamarudin et al.

Published by Auris Reference Limited
www.aurisreference.com

United Kingdom

Handbook of Power Electronics

ISBN: 978-1-78154-926-1

British Library Cataloguing in Publication Data
A CIP record for this book is available from the British Library

Printed in the United Kingdom

Exclusively distributed by CBS Publishers & Distributors Pvt. Ltd.

Sales & Distribution Rights only for India, Pakistan, Bangladesh, Sri Lanka, Nepal and Bhutan.This book is not to be sold outside these territories.

Contents

List of Abbreviations

ABB	Asea Brown Boveri
ARCPI	Auxiliary resonant commutated pole inverter
CENELEC	Comité Européen de Normalisation Electrotechnique
CF-MR	Constant-frequency multi-resonant
CCM	Continuous conduction mode
CSI	Current source inverter
DBFC	Direct borohydride fuel cell
DMFC	Direct methanol fuel cell
DCM	Discontinuous current mode
EMI	Electromagnetic interference
FBSOA	Forward bias safe operating area
GTO	Gate turn-off
GBFC	Glucose biofuel cell
HIPWM	Harmonic injection PWM
HVDC	High voltage direct current
IGBT	Insulated gate bipolar transistor
IPDM	Integral pulse density modulation
IEC	International Electrical Commission
LRB	Laboratory of Bioclinical Radiopharmaceutics
LED	Light-emitting diode
LRC	Load resonant converter
MMF	Magnetomotive force
NPT	Nonpunch-through
OCV	Open-circuit voltage
PRC	Parallel resonant converter
PWM	Pulsewidth-modulated
QRC	Quasi-resonant converter
RBSOA	Reverse-bias safe operating area
rms	Root mean square
SRC	Series resonant converter
SPRC	Series–Parallel Resonant Converter
SCR	Silicon-controlled rectifier
SEPIC	Single-ended primary inductance converter
SFM	Solderable Front Metal
SOA	Square safe operating area
SynRM	Synchronous reluctance motor
TEM	Two-edge-modulation
UPS	Uninterruptible power supply
VCO	Voltage controlled oscillator
VRM	Voltage regulator module
VSI	Voltage source inverter

ZACE	Zero-average-current error
ZC-MR	Zero-current multi-resonant
ZCS	Zero-current switching

List of Contributors

M. S. Ali
Fuel Cell Institute, Universiti Kebangsaan Malaysia (UKM), 43600 Bangi, Selangor, Malaysia

S. K. Kamarudin
Fuel Cell Institute, Universiti Kebangsaan Malaysia (UKM), 43600 Bangi, Selangor, Malaysia
Department of Chemical and Process Engineering, Universiti Kebangsaan Malaysia (UKM), 43600 Bangi, Selangor, Malaysia

M. S. Masdar
Department of Chemical and Process Engineering, Universiti Kebangsaan Malaysia (UKM), 43600 Bangi, Selangor, Malaysia

A. Mohamed
Department of Electrical, Electronic and System, Universiti Kebangsaan Malaysia (UKM), 43600 Bangi, Selangor, Malaysia

Ioannis Ch. Proimadis
Electromechanical Energy Conversion, Department of Electrical and Computer Engineering, University of Patras, 26504, Rio, Patras, Greece

Dionysios V. Spyropoulos
Electromechanical Energy Conversion, Department of Electrical and Computer Engineering, University of Patras, 26504, Rio, Patras, Greece

Epaminondas D. Mitronikas
Electromechanical Energy Conversion, Department of Electrical and Computer Engineering, University of Patras, 26504, Rio, Patras, Greece

A. Zebda
Univ Grenoble 1, CNRS, De´partement de Chimie Moleculaire, UMR-5250, ICMG FR-2607, BP-53, 38041 Grenoble Cedex 9, France
UJF-Grenoble 1/CNRS/TIMC-IMAG UMR 5525, Grenoble, F-38041, France

S. Cosnier
Univ Grenoble 1, CNRS, De´partement de Chimie Moleculaire, UMR-5250, ICMG FR-2607, BP-53, 38041 Grenoble Cedex 9, France

J.-P. Alcaraz
UJF-Grenoble 1/CNRS/TIMC-IMAG UMR 5525, Grenoble, F-38041, France

M. Holzinger
Univ Grenoble 1, CNRS, Département de Chimie Moleculaire, UMR-5250, ICMG FR-2607, BP-53, 38041 Grenoble Cedex 9, France

A. Le Goff
Univ Grenoble 1, CNRS, Département de Chimie Moleculaire, UMR-5250, ICMG FR-2607, BP-53, 38041 Grenoble Cedex 9, France

C. Gondran
Univ Grenoble 1, CNRS, Département de Chimie Moleculaire, UMR-5250, ICMG FR-2607, BP-53, 38041 Grenoble Cedex 9, France

F. Boucher
UJF-Grenoble 1/CNRS/TIMC-IMAG UMR 5525, Grenoble, F-38041, France

F. Giroud
Univ Grenoble 1, CNRS, Département de Chimie Moleculaire, UMR-5250, ICMG FR-2607, BP-53, 38041 Grenoble Cedex 9, France

K. Gorgy
Univ Grenoble 1, CNRS, Département de Chimie Moleculaire, UMR-5250, ICMG FR-2607, BP-53, 38041 Grenoble Cedex 9, France

H. Lamraoui
UroMems - 46 av. Félix Viallet - 38031 GRENOBLE cedex, France.

P. Cinquin
UJF-Grenoble 1/CNRS/TIMC-IMAG UMR 5525, Grenoble, F-38041, France

C. Carretero
Dep. Ingeniería Electrónica y Comunicaciones, Universidad de Zaragoza, María de Luna, 1, 50018 Zaragoza, Spain

J. Acero
Dep. Ingeniería Electrónica y Comunicaciones, Universidad de Zaragoza, María de Luna, 1, 50018 Zaragoza, Spain

R. Alonso
Dep. Física Aplicada, Universidad de Zaragoza, Pedro Cerbuna, 12, 50009 Zaragoza, Spain

Preface

Power electronics is the study of switching electronic circuits in order to control the flow of electrical energy. Power electronics is the technology behind switching power supplies, power converters, power inverters, motor drives, and motor soft starters. The text *Handbook of Power Electronics* focuses on the fundamental principles, models, and technical requirements needed for designing practical power electronic systems. First chapter presents a review of power electronics applications in fuel cell systems, which include various topology combinations of DC converters and AC inverters and which are primarily used in fuel cell systems for portable or stand-alone applications. An alternative for all-electric ships applications has been discussed in second chapter. Third chapter deals with single glucose biofuel cells implanted in rats power electronic devices. The power losses in the winding of magnetic devices used in power systems made with multi-stranded litz-wire are analyzed in fourth chapter. The fundamentals of power electronics have been focused in fifth chapter. Sixth chapter discusses on three-phase controlled rectifiers. Resonant and soft-switching techniques in power electronics have been described in last chapter.

Chapter 1

AN OVERVIEW OF POWER ELECTRONICS APPLICATIONS IN FUEL CELL SYSTEMS: DC AND AC CONVERTERS

M. S. Ali[1], S. K. Kamarudin[1,2], M. S. Masdar[2], and A. Mohamed[3]

[1]Fuel Cell Institute, Universiti Kebangsaan Malaysia (UKM), 43600 Bangi, Selangor, Malaysia

[2]Department of Chemical and Process Engineering, Universiti Kebangsaan Malaysia (UKM), 43600 Bangi, Selangor, Malaysia

[3]Department of Electrical, Electronic and System, Universiti Kebangsaan Malaysia (UKM), 43600 Bangi, Selangor, Malaysia

ABSTRACT

Power electronics and fuel cell technologies play an important role in the field of renewable energy. The demand for fuel cells will increase as fuel cells become the main power source for portable applications. In this application, a high-efficiency converter is an essential requirement and a key parameter of the overall system. This is because the size, cost, efficiency, and reliability of the overall system for portable applications primarily depend on the converter. Therefore, the selection of an appropriate converter topology is an important and fundamental aspect of designing a fuel cell system for portable applications as the converter alone plays a major role in determining the overall performance of the system. This paper presents a review of power electronics applications in fuel cell systems, which include various topology combinations of DC converters and AC inverters and which are primarily used in fuel cell systems for portable or stand-alone applications. This paper also reviews the switching techniques used in power conditioning for fuel cell systems. Finally, this paper addresses the current problem encountered with DC converters and AC inverter.

INTRODUCTION

Renewable energy systems offer environmental and economic advantages in producing energy compared with the conventional fossil fuel systems. Among all types of green energy applications, fuel cells are the most popular because they can provide a continuous power supply throughout all seasons as long as fuel is provided. In comparison, other types of green energy such as solar or wind energy are dependent on the weather conditions. However, the renewable energy supplied by fuel cells has a low-voltage output characteristic, and for any potential practical application, a high step-up DC-DC converter is required [1, 2]. Every source of energy normally comes with a set of challenges in terms of efficiency optimisation, ramifications on the environment, and utilisation. The characteristic electrical output of fuel cells has some drastic shortcomings and a low output voltage that decreases as the load current increases. Fuel cell stacks are also unable to meet transient electrical power demands to the load. However, the low emission and high efficiency of the fuel cell make them a favourable choice for energy sources in portable applications [3, 4].

Implementing power electronics applications in fuel cell systems is a solution that allows fuel cell technology to be used in any application. A fuel cell system can be used in any application with the right selection of power electronics circuits. Using different switching components and switching topologies in a power electronics circuit will yield different results and efficiencies. With the advances in technology made in the semiconductor industry, hard switching power electronics components were added to improve reliability and efficiency. The size of switching components is becoming smaller, which makes them cost-effective. However, there is still a large amount of electrical noise present. The idea behind using a soft-switching method is purposely to reduce noise and switching losses. In addition, switching losses can be reduced by implementing zero-voltage switches (ZVSs) or zero-current switches (ZCSs) [5]. This paper presents a review of power electronics applications in fuel cell systems, and it includes various topologies combinations of DC converters and AC inverters, which are primarily used in fuel cell systems for portable or standalone applications. This paper also reviews the switching techniques used in power conditioning for fuel cell systems.

CURRENT TECHNOLOGY BEHIND THE MAIN TOPOLOGY OF DC CONVERTERS

Several DC converters are available that can increase or decrease the magnitude of the DC voltage and/or invert its polarity. The switch is realised using a power MOSFET and diode; however, other semiconductor switchers, such as IGBTs, BJTs, or thyristors, can be used according to the application [7]. Besides, many semiconductor switchers offered which greatly reduce the switching losses are available in market such as ultra-fast 1200 V IGBTs which can reduce switching and conduction losses with renewable energy technologies focusing on delivering the highest efficiency and reliability. These 1200 V Trench IGBTs are designed to meet the most demanding of any system performance requirement which has been optimized for lowest switching losses and smoothest turn-off in higher frequency. New approaches of the Solderable Front Metal (SFM) technology greatly extend power cycling capability while dual-sided cooling further reduces power dissipation to provide a highly efficient solution [8]. Another solution is Silicon Carbide (SiC) known as SiC-based power electronics which can reduces the size and switch losses in power system by 50% focussing especially on the high power electronics application such as power utilities, smart grids, high-power industrial drive, and renewable energy panel [9].

DC-DC Boost Converter

Basically, a boost converter consists of switching element (M), a diode (D), an inductor (L), and a switching controller, as shown in Figure 1. The switching element is switched between the "on" and "off" state by the controller to boost the input voltage to the desired output voltage. During the "on" state of the switching element, electrical energy is stored in the inductor, and then the capacitor supplies current to the load and the diode with a reverse bias. When in the "off" state, stored energy is transferred to the load and capacitor through the diode [6]. The boost converter operates in one of two modes, continuous-conduction mode or discontinuous-conduction mode, which is characterised by the current waveform of the inductor [10]. The inductor current is greater than zero all the time when in continuous-conduction mode, and the inductor current falls to zero after each switching cycle when in discontinuous-conduction mode [6].

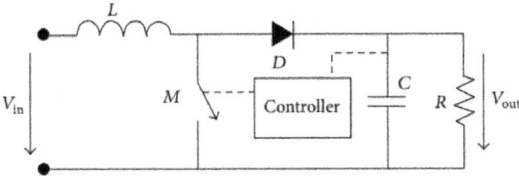

Figure 1: DC-DC boost converter topology, Brey et al. [6].

The boost converter has characteristics such as a continuous input current, a low input ripple current, and good clamping of output diodes requiring less

than half of the challenging task as it invariably involves the analysis of various voltage ratings of rectifier diodes, which are ideal for fuel cell applications, and many attempts have been made to improve the efficiency of the boost topology at low input voltages, making it the ideal choice for fuel cell applications. These improvements include optimising the transformer design to achieve very low leakage inductances, taking advantage of modern power electronics switching, and eliminating the need for a voltage clamp circuit [12–19].

Yang et al. [20] proved that a new quadratic boost converter with an additional capacitor-inductor-diode (CLD) cell has better characteristics than the conventional quadratic boost converter and that the topology is suitable for extreme high-voltage step-up ratio applications. Yao et al. [21] analysed the full-bridge (FB) boost converter that combines the two-edge-modulation (TEM) scheme and the FB cell that is leading-edge modulated with a three-mode PS-TEM control scheme used to improve the efficiency and reliability. Apparently, they proved that the operation of the FB-boost converter and three-mode dual-frequency PS-TEM control are more effective, while Hwu and Yau [22] demonstrated the combination of a KY boost with a traditional synchronously rectified (SR) boost converter for low-ripple applications and found that the efficiency is 90% or more above the half-load point.

Park et al. [24] and Bo et al. [25] proposed a new soft-switching boost converter using a soft-switching method with a resonant inductor and capacitor, an auxiliary switch, and diodes and proved that the converter reduces switching losses more than a conventional hard-switching converter and that the efficiency increases approximately 96%. Ivanovic et al. [26] developed new control algorithms for higher converter efficiency that require loss model parameters that can be used in the algorithm to improve the efficiency of the boost converter.

DC-DC Buck Converter

The buck converter is used to reduce the DC voltage and has a conversion ratio of M(D)=D. It is widely used because of its simple topology, which is characterised by a low number of components, low control complexity, and no isolation. In the conventional buck converter, only a single active switch is used, and the maximum voltage applied across the terminals of the semiconductors equals the input voltage obtained using the hard switching technique. However, this conventional method produces low efficiency because of the high conduction losses due to high-voltage-rated devices and high switching losses. Buck converters can be used in low-power-range regulators and very high range step-down converters [27, 28]. Rodrigues et al. [29] presented a study of a DC-DC buck converter with three-level buck clamping, zero-voltage switching (ZVS), active clamping, and constant-frequency pulse width modulation (PWM) and proved that the voltages across the switch are 50% lower compared with a two-level ZVS buck-buck converter. Chen et al. [30] proposed a new single-inductor quadratic converter using average-current-mode control without slope compensation, which minimises several power management problems, such as efficiency, EMI, size, transient response, design complexity, and cost. Jahanmahin et al. [31] proposed an improved configuration for DC-DC buck and boost converters, which is a novel method for increasing output power by utilising two storage elements and reducing the output ripple voltage for buck and boost converters.

DC-DC Buck-Boost Converter

The basic circuit of a buck-boost converter consists of a switching element, inductor, diode, and capacitor. The difference between a buck-boost and a boost converter is the arrangement of the switching element placed before the inductor, as shown in Figure 2. The buck-boost topology can produce an output voltage that is equal to, less than, or greater than the input voltage. The buck-boost topology is suitable for portable applications, which require a wide range of output levels, and it is an attractive choice when a large current is supplied [6]. The output voltage is equal to the input voltage when the duty cycle is 0.5. When the duty cycle is less than 0.5, the buck-boost converter operates in buck mode, causing the output voltage to be lower than the input voltage. To make the buck-boost converter operate in boost mode and cause the output voltage to be higher than the input voltage, the duty cycle must be greater than 0.5. This relationship is shown in Figure 3 [6].

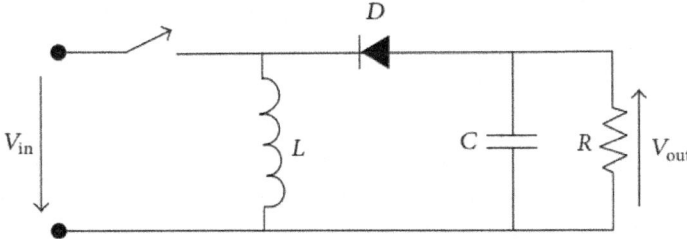

Figure 2: DC-DC buck-boost converter topology, Brey et al. [6].

Chen et al. [32] proposed a buck-boost PWM converter having two independently controlled switches that can work as a boost or as a buck converter depending on the input-output conditions; this approach puts lower stresses on the components. Hwang et al. [33] proposed a low-voltage positive buck-boost converter using an average current controlled with a simple compensation design. This approach can reduce some power management problems, such as size, cost, design complexity, and a simple compensation design, and it provides regulated output with a maximum efficiency of 72% at a switching frequency of 1 MHz. Boopathy and Bhoopathy Bagan [34] presented a novel method of implementing a real-time buck-boost converter with an improved transient response for low-power portable applications and significantly improved the efficiency from 16% and 19%. Hwu and Peng [35] proposed a novel buck-boost converter combining KY and buck converters, which can solve the problem of voltage bucking of the KY converter and increase the application capability of the KY converter with an efficiency above the DC load current of 0.25 A of 88% or more.

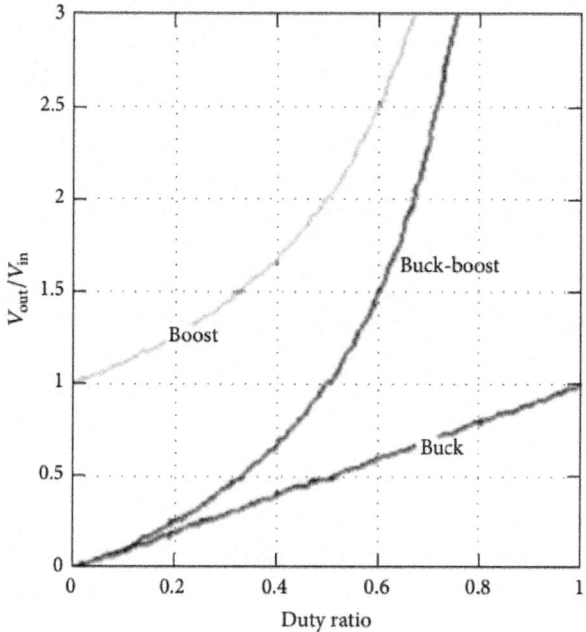

Figure 3: Comparison of duty ratio in buck, boost, and buck-boost converter, Brey et al. [6].

DC-DC Cuk Converter

The Cuk converter contains an inductor in series with the converter input and output port. The switch network alternately connects a capacitor to the input and output inductors. The conversion ratio M(D) is identical to that of the buck-boost converter. Hence, the converter also inverts the voltage polarity, while either increasing or decreasing the voltage magnitude. The Cuk converter is a modified boost-buck converter and can be used either to step up or step down the output voltage with respect to the input. It produces a negative output voltage from a positive input voltage, and its advantage is the presence of both input and output inductors. These inductors lower the current ripple on the input source and the load. Furthermore, the Cuk converter also has a higher efficiency, reduced EMI generation, and a better dynamic response [36]. Lin et al. [37] designed and demonstrated an active snubber zero-voltage switching Cuk converter for achieving parallel operation and balanced current sharing.

DC-DC SEPIC Converter

The single-ended primary inductance converter (SEPIC) can either increase or decrease the voltage magnitude. However, it does not invert the polarity. The conversion ratio is $M(D)=D/(D-1)$. The SEPIC has the features of the buck-boost operating mode, no polarity inversion, low input current pulsation, and a wide input voltage range. It also has the following advantages, which are applied to the electricity generation system of the fuel cell: (i) It is a converter that can operate under boost or buck situations, and there is greater elasticity for assisting the design of the auxiliary source. (ii) Compared with other common boost or buck converters, the SEPIC converter has no issues with an inverse polarity output voltage. (iii) The input terminal of a SEPIC converter contains an inductor, which can reduce the input current pulses and overcome the disadvantage of the electric current pulses of the fuel cell to increase the accuracy of the control [38]. The SEPIC is commonly used in light emitting diode (LED) backlights and photovoltaic applications because it can produce noninverting output and also can operate as a step-up and step-down converter. However, its power conversion efficiency is lower than that of other converters, such as the buck and boost converter, because its extra inductor and capacitor cause additional power losses [39–41]. Song et al. [42] proposed a modified SEPIC design for a step-up and step-down converter having higher power conversion efficiency than the original SEPIC topology in continuous conduction mode (CCM). The input power can be directly delivered to the output with few losses.

APPLICATION OF DC CONVERTERS IN FUEL CELL SYSTEMS

Due to the limitations of fuel cells, which include low voltages, low current densities, and unstable power production, the DC converter has become the most important component in fuel cell systems for portable or stand-alone applications. Normally, a single DMFC can supply only approximately 0.3–0.5 V under loaded conditions. By using a DC converter, these limitations can be solved with a converted voltage source from the fuel cell [1]. Various DC converters have been developed to support fuel cell systems, but the efficiency of a DC converter depends on the conduction and switching losses. If there are fewer conduction and switching losses, then the DC converter operation is more efficient. By reducing the number of components used and their operating ranges, the conduction losses can be effectively reduced [43]. A fuel cell system application is necessary to use a power management circuit to generate the desired load voltage level. Therefore, it may require a nominal supply voltage, which can be above, below, or equal to the fuel cell generating voltage [6]. Therefore, the system designed for the nominal supply voltage will require a converter capable of stepping up or stepping down the fuel cell voltage [44]. There are two converters widely used in fuel cell systems for the single-stage level or for low-voltage portable applications to step up and step down the fuel cell voltage with a high efficiency: boost converters (step-up) and buck converters (step-down).

Single-Stage Topologies

To fulfil the operational requirements of fuel cell systems, the researchers developed various topologies for single-stage conversion either using a DC-DC converter or a DC-AC inverter. The voltage generated from the fuel cell can be converted directly into a regulated DC voltage, or it can be converted directly into an AC voltage as a supply voltage, depending on the AC or DC load. A single-stage topology has reduced component counts and is simple to control.

Brey et al. [6] tested several DC converters in their study on the power conditioning of fuel cell systems in portable applications and claimed that the DC-DC boost converter gives the best performance for a 100 W fuel cell power condition supplying the desired regulated output voltage level and maintaining an input current ripple below 2% of the nominal input current. Nymand et al. [46] presented a comprehensive comparison between buck and boost topologies and claimed that the boost converter topology is more appropriate for low-voltage fuel cell applications. Kui-Jun and Rae-Young [47] proposed a zero-voltage-transition (ZVT) two-inductor boost converter using a single resonant inductor with a maximum efficiency at 92.4% due to the ZVS operation under the V_{in}=30 V condition.

Chang et al. [11] proposed a complete hybrid active DMFC system design by implementing a hybrid between the DMFC stack and a Li-ion battery to share the applied load to exploit the high energy density of the fuel cell and the power density of the battery, as presented in Figure 4. A conventional battery charger hybridisation circuit relies on the battery as a primary power source, as shown in Figure 4(a). The DMFC stack operates only when charging the battery meaning that there are no direct supplies from fuel cell to laptop. The system operates starting with the DMFC stack by charging the battery first, and then the battery will supply power to DC-DC converter to operate the laptop. Then for Figure 4(b), there to ways how the fuel cell powered the laptop either direct supply from fuel cell to DC-DC converter and to the laptop or from fuel cell charging the battery first then battery supply power to DC-DC converter to operate the laptop. These two operation ways occurred simultaneously. Supplied power from battery will become additional current for the system to back up a current to another direct supplied from fuel cell. This additional current goes through three converters: the battery charger, the battery output converter, and the output regulation converter.

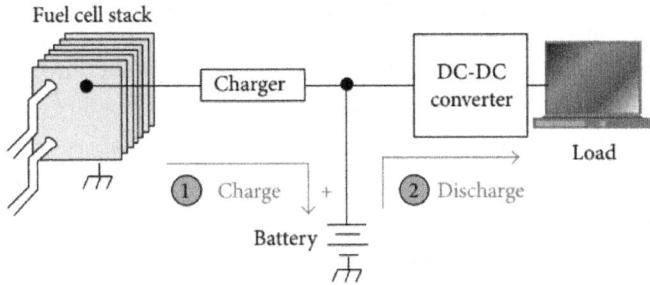

(a) Battery charger architecture

(b) 2-diode and charger architecture

Figure 4: Power control scheme for fuel cell and battery hybrid, Chang et al. [11].

Harfman-Todorovic et al. [49] proposed a hybrid DC-DC converter for a fuel cell-powered laptop computer. However, this power distribution system has problems with a large reduction of voltage, for which it needs a voltage regulator module (VRM). Alternative power distribution architecture with difference interfacing energy sources is presented, and the hybrid converter proved the capability of interfacing a DC-link with multi-input power sources in a 30 W fuel cell.

Multistage Topology

Power conditioning combinations involving two types of DC-DC converters or an AC inverter are recognised as multistage topology. In this topology voltage or current of the fuel cell will convert first to desired value by using DC-DC converter. This DC-DC converter may involve any type of DC-DC converter such as boost, buck-boost, and so on to get desired value of voltage or current becoming input variable to the inverter and is also used to charge ultra-capacitor or battery. Second step in this topology is DC voltage from DC-DC converter inverted to AC voltage. The disadvantage of this topology using more components compared with a single stage will make the power loss in a multistage topology greater than that in a single-stage topology.

Zhu et al. [50] proposed a hybrid energy system topology between DMFC and a supercapacitor in which the DMFC and the super capacitor bank are connected to a common DC voltage bus through a boost converter and a bidirectional converter. A similar topology has been proposed and discussed in detail by others [1, 41,42]. Here, a supercapacitor is used to balance the system's power flow for load voltage regulation by delivering the deficit power during heavy load conditions and absorbing the surplus power from the DMFC during light load conditions while the DMFC is constantly tracked at the MPP by the MPPT controller. The supercapacitor can be charged and discharged rapidly under varying load conditions for an uninterrupted power flow to the load with suitable design of the bidirectional converter.

Kwon et al. [51] proposed a high-efficiency active DMFC system for portable applications. This system used a smart battery to support the fuel cell system that consisted of two buck-boost converters to increase the power conversion efficiency. Alotto et al. [23] claimed that the hybrid sources can combine the high power density of batteries with the high energy density of fuel cells to improve the runtime of portable electronic devices, as shown in Figure 5, in which the synchronous boost converter is connected after the fuel cell to boost the voltage from the fuel cell up to the DC-bus voltage (typically 3.3 V for electronic devices). In a real application, the fuel cell will be connected in series to attain acceptable input voltage levels for a full load (2–2.4 V), while on the battery side, the H-bridge buck-boost converter will be employed to manage charging and discharging the battery. This H-bridge buck-boost converter is emerging with an analogue design, which means this converter can be operated as a buck converter when the battery is fully charged or as a boost converter when the battery is almost discharged.

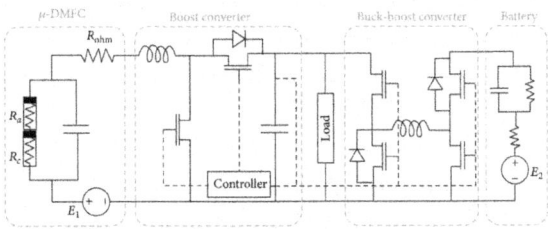

Figure 5: The hybrid power system (small-scale DMFC and Li-ion battery), Alotto et al. [23].

LIMITATIONS OF CURRENT DC CONVERTERS

Boost converters have a voltage collapse point when the output power is too high and the input current becomes excessive, leading to high losses, which in turn reduces the efficiency further, requiring even more input current, causing voltage collapse. There are also large voltage drops across the diode. For low-voltage applications, replacing the diode with a switching element can improve the efficiency of the converter, which results in a synchronous DC-DC boost converter topology, as shown in Figure 6. In this circuit, a time delay must be used to prevent a shoot-through current between turning off M_1 and turning on M_2. In this way, the voltage drop across M_2 is the lowest, and the efficiency can be improved [6]. Two power management schemes were used to achieve a high efficiency within the whole load range: PWM control for the heavy load condition and PFM control for the light load condition [52, 53]. The converter has different power losses (conduction loss and switching loss) under different load conditions.

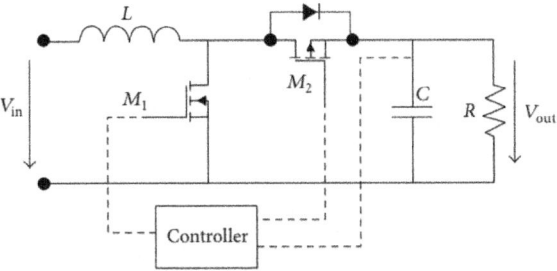

Figure 6: Synchronous DC-DC boost converter topology, Brey et al. [6].

The principal disadvantage of a boost converter is the high switching noise. This noise is generated when turning on and off the switching element, and it deteriorates the quality of the output voltage and degrades the performance of the fuel cell. Several solutions to avoid this problem have been proposed: a snubber circuit can be placed at every switch, additional electromagnetic interface at the input and output of the converter can be incorporated, and the soft switching technique can be used [6].

The buck-boost converter requires a greater duty cycle than the boost converter to boost the input voltage at the same output voltage level. This causes the buck-boost converter to be less efficient compared with the boost converter because higher duty cycles increase conduction losses in the switching element. Another disadvantage is that the obtained output voltage polarity is opposite to the supply voltage, meaning this topology is not suitable for some applications. To solve this problem, a noninverting buck-boost converter can be used [6]. Noninverting buck-boost converter topology involves cascading between a buck converter and a boost converter, as shown in Figure 7. For low-voltage applications, a switch can replace the diode to improve the efficiency. This topology has the same problem with switching noise as the boost converter topology [6].

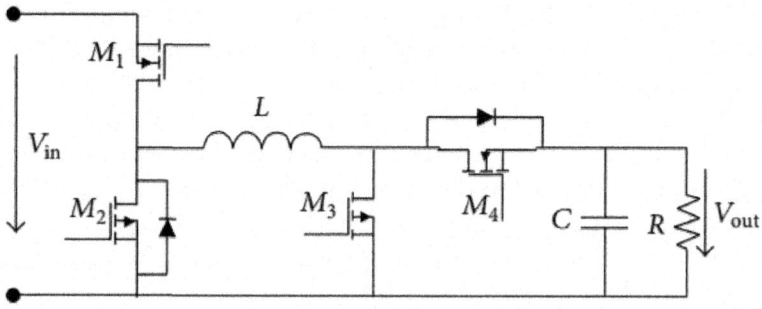

Figure 7: Noninverting buck-boost converter topology, Brey et al. [6].

One of the biggest goals in power conditioning design is to operate the converter with a high switching frequency to minimise both the cost and size of the respective power conditioning and also to optimise the efficiency. Conventional hard-switched pulse-width modulation (PWM) suffers from high switching losses, high device stress, and objectionable EMI when operating at a high switching frequency compared with the soft-switching PWM technique, which is being used for high switching frequency operation with high efficiencies and large power-to-volume ratios [45].

RECENT DEVELOPMENT IN DC-DC CONVERTERS

Saha [45] proposed that the efficient soft-switched boost converter can be increased by using an auxiliary network in addition to the boost inductor L_s, boost switch S_1, and boost diode D_2 as shown in Figure 8. The auxiliary network consists of on-switch element S_2, one diode D_1, two inductors L_1, L_2, and one snubber capacitor C_s. This converter operates under seven modes, and it was determined that the converter can operate with a high efficiency (above 95%) at full load. The output voltage of this converter can be regulated by varying the pulse-width of the main switch S_1. Lin and Dong [48] proposed a new zero-voltage switching DC-DC converter for renewable energy conversion systems based on a boost converter and a voltage-doubler configuration with a coupled inductor to achieve a high step-up voltage conversion ratio, as shown in Figure 9. In this configuration, an active snubber is adopted to clamp the voltage stress of the active switch and to release the energy stored in leakage and magnetising inductances. In conventional boost converter, the adopted converter has a wide turn-off period to achieve high voltage. By adopting an asymmetrical pulse-width modulation to control active switches, leakage inductance and output capacitance of the active switch are resonant in the transition interval. The result from this configuration is that both switchers turn on at zero-voltage switching which overcome the problem of the conventional boost converter with low circuit efficiency and narrow turn-off period. Al-Saffar et al. [54] proposed a new soft-switched pulse-width modulation (PWM) quadratic boost converter that is suitable for application to a wide fluctuating DC input voltage range. The voltage gain of the conventional PWM DC-DC boost converter has limitations for practical applications, even under extreme duty cycle conditions, due to parasitic components [55]. A new soft-switched pulse-width-modulated (PWM) quadratic boost converter that is suitable for applications with a wide fluctuating DC input voltage range is designed for fuel cell systems. The efficiency of this converter is equal to 92.3%. Finally, Delshad and Farzanehfard [56] proposed a new zero-voltage switching current-fed push-pull DC-DC converter. The auxiliary circuit in this method is introduced purposely to provide a zero-voltage switching condition and also to absorb the voltage surge across the switches at the turn-off instance. Therefore, the size and weight of the converter can be reduced, and the efficiency of the converter can be increased. To reduce the size and weight of the converter, high-frequency operation is required to reduce the size of the induction element and other reactive components. This converter controls via PWM with a very simple implementation control circuit. This converter can be operated at higher frequencies, and these frequencies are higher than the frequency of a conventional current-fed converter. In this converter, the clamp circuit not only absorbs the voltage surge across the power switches but also provides soft switching conditions for all semiconductor devices. Because the converter uses only one input filter inductor and it does not use any clamp winding, the cost and size can be reduced compared with conventional converters.

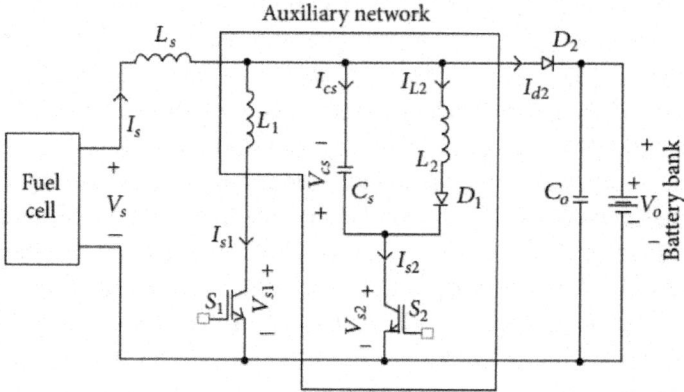

Figure 8: New soft-switched DC-DC boost converter, Saha [45].

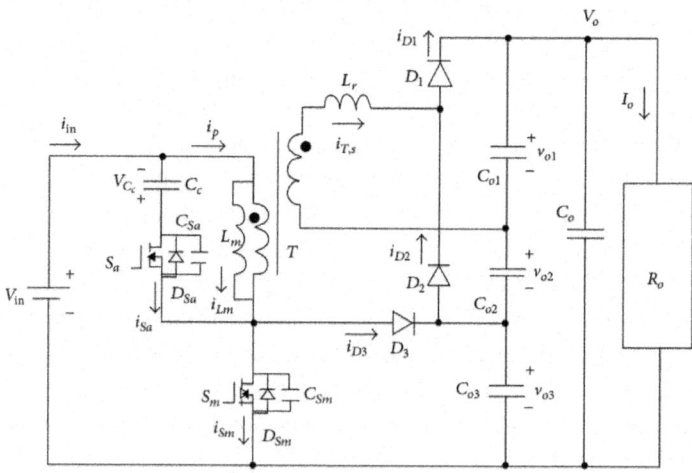

Figure 9: New zero-voltage switching DC-DC converter, Lin and Dong [48].

CONCLUSIONS

The DC converters with various topologies contribute to the use of renewable energy in various applications, especially in portable or stand-alone applications. A review on DC converters shows that they can be used to produce a more efficient conversion of power from the fuel cell to the load. Using a DC converter or a combination of DC converters can address the limitations of fuel cell, which include unregulated voltage, low voltage, low current density, and unstable power. A hybrid DC converter with a battery or a super capacitor or other external supplies can stabilize the power conditioning to balance the excess and insufficient power condition in the fuel cell. This review also shows that the switching technique is the main element of a DC converter. Introduction of the soft-switching operation to DC converters introduces improvements in terms of increased converter power density and converter efficiency. The topology of the DC converter in the power conditioning unit can be divided into single-stage and multistage topologies. In single-stage topologies, the DC converter stands alone, but in multistage topologies, the DC converter is combined with DC converters or with AC inverters.

In conclusion, the design of DC converter topologies is considered important in fuel cell systems. Therefore, more studies on the development of new topologies for DC converters, including new switching techniques, are needed to create a higher efficiency and improve the existing switching technique. For specific applications of portable fuel cell systems, the size and energy density are considered very important. Currently, portable fuel cell focuses on two types of fuel cells that can fulfil the size and energy density requirements: direct methanol fuel cells (DMFCs) and direct borohydride fuel cell (DBFCs). Fuel cells such as DMFCs and DBFCs with improved power converter technologies can be considered as promising alternative energy sources for portable applications.

ACKNOWLEDGMENTS

The authors gratefully acknowledge financial support for this work by the Malaysian Ministry of Science, Technology and Innovation (MOSTI) under Research University Grant nos. UKM-DIP-2012-04 and DPP-2013-113.

REFERENCES

1.   F. Blaabjerg, Z. Chen, and S. B. Kjaer, "Power electronics as efficient interface in dispersed power generation systems," IEEE Transactions on Power Electronics, vol. 19, no. 5, pp. 1184–1194, 2004.

2.   J. M. Carrasco, L. G. Franquelo, J. T. Bialasiewicz et al., "Power electronic systems for the grid integration of renewable energy sources: a survey," IEEE Transactions on Industrial Electronic, vol. 53, no. 4, pp. 1002–1016, 2006. ·

3.   C. Wang, M. H. Nehrir, and S. R. Shaw, "Dynamic models and model

validation for PEM fuel cells using electrical circuits," IEEE Transactions on Energy Conversion, vol. 20, no. 2, pp. 442–451, 2005.

4.   A. Forrai, H. Funato, Y. Yanagita, and Y. Kato, "Fuel-cell parameter estimation and diagnostics," IEEE Transactions on Energy Conversion, vol. 20, no. 3, pp. 668–675, 2005.

5.   S. Wang, Y. Kenarangui, and B. Fahimi, "Impact of Boost converter switching frequency on optimal operation of fuel cell systems," in Proceedings of the IEEE Vehicle Power and Propulsion Conference (VPPC '06), pp. 1–5, September 2006.

6.   J. J. Brey, C. R. Bordallo, J. M. Carrasco, E. Galván, A. Jimenez, and E. Moreno, "Power conditioning of fuel cell systems in portable applications," International Journal of Hydrogen Energy, vol. 32, no. 10-11, pp. 1559–1566, 2007.

7.   R. W. Erickson, Fundamentals of Power Electronics, Chapman and Hall, New York, NY, USA, 1997.

8.   "Power Electronic," http://www.irf.com/pressroom/pressreleases/nr091112.html.

9.   Global Power, http://www.gptechgroup.com/index.php/en/news/214-sic-based-power-electronics-reduces-the-size-and-switch-losses-in-power-systems-by-50.

10. L. Palma, M. H. Todorovic, and P. Enjeti, "Design considerations for a fuel cell powered DC-DC converter for portable applications," in Proceedings of the 21st Annual IEEE Applied Power Electronics Conference and Exposition (APEC '06), pp. 1263–1268, March 2006.

11. N. Chang, J. Seo, D. Shin, and Y. Kim, "Room-temperature fuel cells and their integration into portable and embedded systems," in Proceedings of the 15th Asia and South Pacific Design Automation Conference (ASP-DAC '10), pp. 69–74, Taipei, Taiwan, January 2010. ··

12. W. Li, J. Liu, J. Wu, and X. He, "Design and analysis of isolated ZVT boost converters for high-efficiency and high-step-up applications," IEEE Transactions on Power Electronics, vol. 22, no. 6, pp. 2363–2374, 2007.

13. G. K. Andersen, C. Klumpner, S. B. Kjær, and F. Blaabjerg, "A new power converter for fuel cells with high system efficiency," International Journal of Electronics, vol. 90, no. 11-12, pp. 737–750, 2003.

14. J.-T. Kim, S.-J. Jang, B.-K. Lee, S.-S. Kim, T.-W. Lee, and C.-Y. Won, "An active clamping current-fed half-bridge converter for fuel-cell generation systems," in Proceedings of the IEEE 35th Annual Power Electronics Specialists Conference (PESC '04), pp. 4709–4714, Aachen,

Germany, June 2004.

15. H. Xiao, L. Guo, and S. Xie, "A new ZVS bidirectional DC-DC converter with phase-shift plus PWM control scheme," in Proceedings of the 22nd Annual IEEE Applied Power Electronics Conference and Exposition (APEC '07), pp. 943–948, Anaheim, Calif, USA, March 2007.

16. K. Wang, C. Y. Lin, L. Zhu, D. Qu, F. C. Lee, and J.-S. Lai, "Bi-directional DC to DC converters for fuel cell systems," in Proceedings of the Power Electronics in Transportation, pp. 47–51, Dearborn, Mich, USA, October 1998. ·

17. K. Wang, L. Zhu, D. Qu, H. Odendaal, J. Lai, and F. C. Lee, "Design, implementation, and experimental results of bi-directional full-bridge dc/dc converter with unified soft-switching scheme and soft-starting capability," in Proceedings of the IEEE 31st Annual Power Electronics Specialists Conference (PESC '00), pp. 1058–1063, 2000.

18. L. Zhu, "A novel soft-commutating isolated boost full-bridge ZVS-PWM DC-DC converter for bidirectional high power applications," IEEE Transactions on Power Electronics, vol. 21, no. 2, pp. 422–429, 2006.

19. M. Nymand and M. A. E. Andersen, "A new approach to high efficiency in isolated boost converters for high-power low-voltage fuel cell applications," in Proceedings of the 13th International Power Electronics and Motion Control Conference (EPE-PEMC '08), pp. 127–131, Poznan, Poland, September 2008.

20. P. Yang, J. Xu, G. Zhou, and S. Zhang, "A new quadratic boost converter with high voltage step-up ratio and reduced voltage stress," in Proceedings of the IEEE 7th International Power Electronics and Motion Control Conference (IPEMC '12), pp. 1164–1168, June 2012. · ·

21. C. Yao, X. Ruan, X. Wang, and C. K. Tse, "Isolated buck-boost DC/DC converters suitable for wide input-voltage range," IEEE Transactions on Power Electronics, vol. 26, no. 9, pp. 2599–2613, 2011.

22. K. I. Hwu and Y. T. Yau, "A KY boost converter," IEEE Transactions on Power Electronics, vol. 25, no. 11, pp. 2699–2703, 2010.

23. P. Alotto, M. Guarnieri, and F. Moro, "Modeling and control of fuel cell-battery hybrid power systems for portable electronics," in Proceedings of the 43rd International Universities Power Engineering Conference (UPEC '08), Padova, Italy, September 2008.

24. S.-H. Park, S.-R. Park, J.-S. Yu, Y.-C. Jung, and C.-Y. Won, "Analysis and design of a soft-switching boost converter with an HI-bridge auxiliary resonant circuit," IEEE Transactions on Power Electronics, vol. 25, no. 8, pp. 2142–2149, 2010.

25. C. Bo, W. Congling, and H. Yao, "Research on soft switching boost converter," in Proceedings of the 2nd International Conference on Digital Manufacturing and Automation (ICDMA '11), pp. 1015–1018, IEEE, August 2011.

26. Z. Ivanovic, B. Blanusa, and M. Knezic, "Power loss model for efficiency improvement of boost converter," in Proceedings of the 23rd International Symposium on Information, Communication and Automation Technologies (ICAT '11), pp. 1–6, October 2011. · ·

27. W. Huang, "A new control for multi-phase buck converter with fast transient response," in Proceedings of the 16th Annual IEEE Applied Power Electronics Conference (APEC '01), pp. 273–279, March 2001.

28. T. Fuse, M. Ohta, M. Tokumasu, H. Fujii, S. Kawanaka, and A. Kameyama, "A 0.5-V power-supply scheme for low-power system LSIs using multi-V_{th} SOI CMOS technology," IEEE Journal of Solid-State Circuits, vol. 38, no. 2, pp. 303–311, 2003.

29. J. P. Rodrigues, S. A. Mussa, M. L. Heldwein, and A. J. Perin, "Three-level ZVS active clamping PWM for the DC-DC buck converter," IEEE Transactions on Power Electronics, vol. 24, no. 10, pp. 2249–2258, 2009.

30. J. J. Chen, B. H. Hwang, C. M. Kung, W. Y. Tai, and Y. S. Hwang, "A new single-inductor quadratic buck converter using average-current-mode control without slope-compensation," in Proceedings of the 5th IEEE Conference on Industrial Electronics and Applications (ICIEA ‹10), pp. 1082–1087, June 2010.

31. M. Jahanmahin, A. Hajihosseinlu, E. Afjei, and M. Mesbah, "Improved configurations for Dc to Dc buck and boost converters," in Proceedings of the 3rd Power Electronics and Drive Systems Technology (PEDSTC ‹12), vol. 2, pp. 372–378, Tehran, Iran, February 2012.

32. J. Chen, D. Maksimović, and R. Erickson, "Buck-boost PWM converters having two independently controlled switches," in Proceedings of the IEEE 32nd Annual Power Electronics Specialists Conference, pp. 736–741, June 2001.

33. B.-H. Hwang, B.-N. Sheen, J.-J. Chen, Y.-S. Hwang, and C.-C. Yu, "A low-voltage positive buck-boost converter using average-current- controlled techniques," in Proceedings of the IEEE International Symposium on Circuits and Systems (ISCAS ‹12), pp. 2255–2258, Seoul, Republic of Korea, May 2012. · ·

34. K. Boopathy and K. Bhoopathy Bagan, "A novel method of implementing real-time buck boost converter with improved transient response for low power applications," in Proceedings of the IEEE Symposium on Industrial

Electronics and Applications (ISIEA ‹11), pp. 155–160, September 2011.

35. K. I. Hwu and T. J. Peng, "A novel buck-boost converter combining KY and buck converters," IEEE Transactions on Power Electronics, vol. 27, no. 5, pp. 2236–2241, 2012.

36. R. W. Erickson and D. Maksimovic, Fundamentals of Power Electronics, Kluwer Academic Publishers, Norwell, Mass, USA, 2nd edition, 2001.

37. B. R. Lin, C. Huang, and H. K. Chiang, "Analysis, design and implementation of an active snubber zero-voltage switching cuk converter," Power Electronics, IET, vol. 1, pp. 50–61, 2008.

38. A. C.-C. Hua and B. C.-Y. Tsai, "Design of a wide input range DC/DC converter based on SEPIC topology for fuel cell power conversion," in Proceedings of the International Power Electronics Conference (IPEC ‹10), pp. 311–316, June 2010.

39. H. S.-H. Chung, K. K. Tse, S. Y. Ron Hui, C. M. Mok, and M. T. Ho, "A novel maximum power point tracking technique for solar panels using a SEPIC or Cuk converter," IEEE Transactions on Power Electronics, vol. 18, no. 3, pp. 717–724, 2003.

40. N. Mohan, T. M. Undeland, and W. P. Robbins, Power Electronics: Converters, Applications, and Design, John Wiley & Sons, 3rd edition, 2002.

41. R. W. Erickson and D. Maksimovic, Fundamental of Power Electronics, Kluwer Academic, Boston, Mass, USA, 2nd edition, 2001.

42. M. S. Song, E. S. Oh, and B. K. Kang, "Modified SEPIC having enhanced power conversion efficiency,"Electronics Letters, vol. 48, no. 18, pp. 1151–1153, 2012. · ·

43. N. Mohan, T. M. Undeland, and W. P. Robbins, "Power electronic converter," in Application and Design, John Wiley & Sons, 3rd edition, 2001.

44. EG&G Technical Services and Science Application International Corporation, Fuel Cell Handbook, U.S. Department of Energy, Office of Fossil Energy, National Energy Technology Laboratory, Morgantown, WVa, USA, 6th edition, 2002.

45. S. S. Saha, "Efficient soft-switched boost converter for fuel cell applications," International Journal of Hydrogen Energy, vol. 36, no. 2, pp. 1710–1719, 2011.

46. M. Nymand, R. Tranberg, M. E. Madsen, U. K. Madawala, and M. A. E. Andersen, "What is the best converter for low voltage fuel cell applications—a buck or boost?" in Proceedings of the 35th Annual

Conference of the IEEE Industrial Electronics Society (IECON ‹09), pp. 962–967, November 2009.

47. L. Kui-Jun and K. Rae-Young, "Nonisolated ZVT two-inductor boost converter with a single resonant inductor for high step-up applications," in Proceedings of the IEEE 8th International Conference on Power Electronics and ECCE Asia (ICPE & ECCE ‹11), 2011.

48. B. R. Lin and J. Y. Dong, "New zero-voltage switching DC-DC converter for renewable energy conversion systems," IET Power Electronics, vol. 5, no. 4, pp. 393–400, 2012.

49. M. Harfman-Todorovic, L. Palma, and P. Enjeti, "A hybrid DC-DC converter for Fuel cells powered laptop computers," in Proceedings of the 37th IEEE Power Electronics Specialists Conference (PESC›06), June 2006.

50. G. R. Zhu, K. H. Loo, Y. M. Lai, and C. K. Tse, "Quasi-maximum efficiency point tracking for direct methanol fuel cell in DMFC/supercapacitor hybrid energy system," IEEE Transactions on Energy Conversion, vol. 27, no. 3, pp. 561–571, 2012.

51. J.-M. Kwon, Y.-J. Kim, and H.-J. Cho, "High-efficiency active DMFC system for portable applications,"IEEE Transactions on Power Electronics, vol. 26, no. 8, pp. 2201–2209, 2011.

52. B. Sahu and G. A. Rincon-Mora, "A low voltage, dynamic, noninverting, synchronous buck-boost converter for portable applications," IEEE Transactions on Power Electronics, vol. 19, no. 2, pp. 443–452, 2004.

53. X. Duan, H. Deng, N. X. Sun, A. Q. Huang, and D. Y. Chen, "A high performance integrated boost DC-DC converter for portable power supply," in Proceedings of the 19th Annual IEEE Applied Power Electronics Conference and Exposition (APEC ‹04), pp. 1039–1044, February 2004.

54. M. A. Al-Saffar, E. H. Ismail, and A. J. Sabzali, "High efficiency quadratic boost converter," in Proceedings of the 27th Annual IEEE Applied Power Electronics Conference and Exposition (APEC ‹12), pp. 1245–1252, February 2012.

55. R. W. Erickson and D. Maksimovic, Fundamental of Power Electronics, Kluwer Academic Publishers, Boston, Mass, USA, 2nd edition, 2001.

56. M. Delshad and H. Farzanehfard, "A new soft switched push pull current fed converter for fuel cell applications," Energy Conversion and Management, vol. 52, no. 2, pp. 917–923, 2011.

Chapter 2

AN ALTERNATIVE FOR ALL-ELECTRIC SHIPS APPLICATIONS: THE SYNCHRONOUS RELUCTANCE MOTOR

Ioannis Ch. Proimadis, Dionysios V. Spyropoulos, and Epaminondas D. Mitronikas

Electromechanical Energy Conversion, Department of Electrical and Computer Engineering, University of Patras, 26504, Rio, Patras, Greece

ABSTRACT

The three-phase synchronous reluctance motor (SynRM) is presented as a possible alternative in all-electric ship applications. The basic features of this motor with regard to the other types of motors are shown. The structure of the motor and specifically the structure of its rotor are analyzed, while the basic operating principles are presented and references on commonly used control strategies are made. In this paper, a demonstration of a reluctance motor fed by a voltage source inverter (VSI) takes place. To demonstrate the operation of the motor fed by a VSI, an example using a scalar control method is implemented, where harmonic injection PWM (HIPWM) is used to drive the VSI. Experimental results on a commercially available motor are shown, focusing on the harmonic content of the current.

INTRODUCTION

Sea transportation plays a crucial role in the development of human civilization for more than 5,000 years [1]. This period has been characterized by a constant effort to design and construct faster, more comfortable, and more reliable ships. However, the most significant step towards the modern sailing has been made during the industrial revolution, as it led to drastic changes in the design and operation of the ships, one of which was the use of fossil-fuel-powered engines for their propulsion.

In modern times, the constant progress in electric and electronic systems had also an important effect on ships and led to the first commercial adoption of electricity-driven technology by the cruise ship industry in the late 1980s [2]. In the latest years, an intensive discussion has emerged around the adoption of the pioneering all-electric ships for the industry [3], since they have more advantages than the mechanical transmission-based ships [4]. This discussion subsequently reinforced the interest in high-end power electronics and electric machines that could be used.

Propulsion systems consume the major part of the total energy in conventional ships [5], while in military ships a considerable proportion of this energy is used to power weapon system equipment. Moreover, new, more advanced, and more automated weapon systems are being developed, which will need much greater amounts of energy to operate [6].

From the above, it is obvious that the need for more efficient electric machines in the future ships will not be restricted to those used for the propulsion systems. In every ship, there is a great need for middle- and low-power motors. These are used in a large variety of applications, such as winches, hoists, and pumps for water or fuels. It is also a fact that the navy is considering the replacement of hydraulic systems and other traditionally mechanical systems by power-electronics-driven electric motors. Also, electric motors are used in combustion air fans, refrigeration plants and ventilation systems. While electric propulsion motors should have a very small weight-to-power ratio, for the large amount of supplementary drive systems needed it is required that an optimum solution concerning efficiency, cost, and performance should be made. It is possible that this optimum solution can be given in the near future by the reluctance motor.

Reluctance motors are considered as a subcategory of synchronous motors with a salient-pole rotor. They differ from other types of motors, since their rotor does not have windings and is constructed from inexpensive materials like iron (instead of expensive magnet, like permanent magnet synchronous motors); thus, there is no concern with demagnetization [7].

Although the reluctance motor has a long history [9], its applications were limited until the last decade, mainly due to lower efficiency than permanent magnet synchronous machines and control issues. However, the development of power semiconductor technology, computers, and microprocessors has revived the interest in these machines in the last years.

Early evolution of reluctance motors led to two different types: switched and synchronous reluctance motors. Although a significant progress has taken place concerning design and control of switched reluctance machines [10], progress regarding synchronous reluctance machines (SynRMs) was not so intensive in the past. However, significant research has been conducted in the recent years, since converter-fed SynRMs show some special features and advantages compared to other types of motors [11–13]:(i)their stator is identical to that of an induction motor, so the stator can be constructed in the already available assembly line.(ii)They can achieve high power densities in regard to their size.(iii)They can function without a starting cage, since they can start in synchronous mode.(iv)Compared to the induction motor, the SynRMs achieve high torque output in regard to iron losses and higher efficiency [11, 12].

The exploitation of these characteristics of SynRM in applications is still on the way, with many of these focusing on hybrid electric vehicles [13]. More extended research has been conducted in this field with the use of switched reluctance and permanent magnet motors [14, 15]; however, the aforementioned features of SynRM indicate that these could be used in the future, in applications that currently use these two motors.

Moreover, the most appropriate converter topologies should be examined, which will lead to adequate performance of the reluctance machine. As far as converter topologies for use in ship applications are concerned, Controlled Rectifier, Load Commutated Inverter, Cycloconverter, and Voltage Source Inverter are the most popular ones [1, 16–22]. The latter is the most prominent topology for the future of all-electric ships and also the most appropriate topology for the control of SynRM. Voltage source inverter (VSI) uses the pulse width modulation (PWM) technique and it is based on power devices with controlled turn-off, such as IGBTs.

In this paper, the performance of SynRM will be studied through the utilization of a simple PWM control method. This control scheme will be applied on a mass-production motor, and attention to the harmonic content of the current which is the main cause for torque ripples will be paid.

ANALYSIS OF THE SYNCHRONOUS RELUCTANCE MO-TOR

Structure of Synchronous Reluctance Motor

As it has been already mentioned, the stator of a SynRM is identical to the stator of an induction machine. The construction of the rotor, however, is more complex. In general, the main concept for the construction of the SynRM is the optimization of the rotor in order to maximize the produced (reluctance) torque. The evolution of the SynRM rotors has led to diverse structures, as it is shown in the Figure 1, but a common attribute of all these is the high rotor field anisotropy [8, 23].

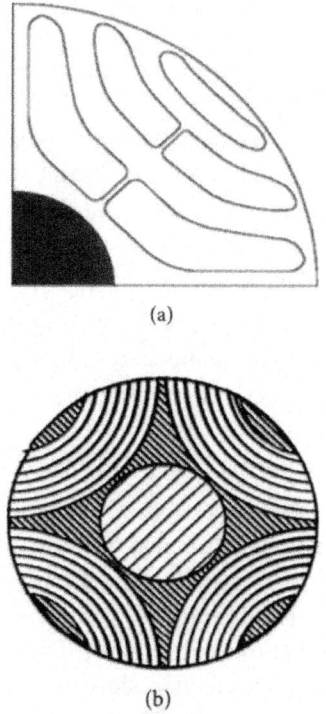

(a)

(b)

Figure 1: Two different rotor designs (transverse and axially laminated rotor).

The individuality of the SynRM's rotor has led to limited adoptions from the industry. This is more characteristic for the axially laminated rotor, although the high L_d/L_q ratio that can be achieved attracted the interest of many researches. On the other hand, transverse laminated rotors are more appropriate for mass production, are rotor core-loss free, and the air-gap harmonics can be subdued through a sophisticated design. More general, the industrial manufacturing of SynRM has to balance between easy, cheap construction of the rotors and efficiency (low current harmonics, reduction of core loss, and low torque ripple). As this paper is focused on practical applications of SynRM, a mass-production motor will be investigated in Section 4.

Basic Operating Principles

As it is well known, the stator windings generate a spatial sinusoidally distributed magnetomotive force (MMF) in the air-gap between stator and rotor. The rotor of SynRM is constructed by ferromagnetic steel and nonmagnetic laminations. Despite the absence of field winding in the rotor, the rotor due to the presence of flux paths (Figure 1) distorts the flux density distribution that is produced by the sinusoidally distributed mmf. A simplified interpretation of the produced torque that forces the rotor to rotate at the synchronous speed (the speed of the field in the airgap) lies in the fact that the flux density tends to align the domains in the ferromagnetic rotor. Due to the rotor field distribution, this is equal to the alignment of the least-reluctance axis with the mmf, or, according to Figure 2, the alignment of d-axis with F vector. However, because of the equal rotation speed of both vectors, the distance between d-axis and F remains constant.

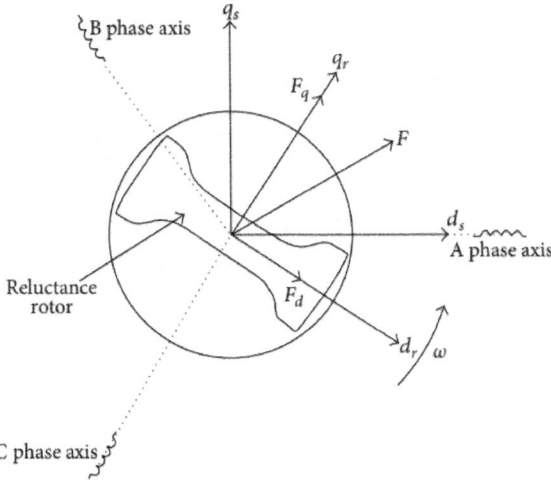

Figure 2: d-q analysis of SynRM [8].

Mathematical Equations

In order to have a comprehensive representation of SynRM, it is significant to present the equivalent circuits (Figure 3) and the mathematical model of the machine.

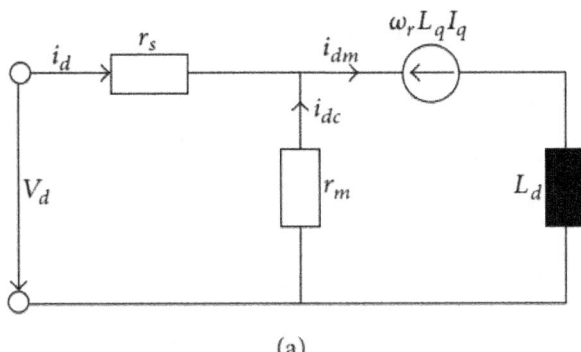

(a)

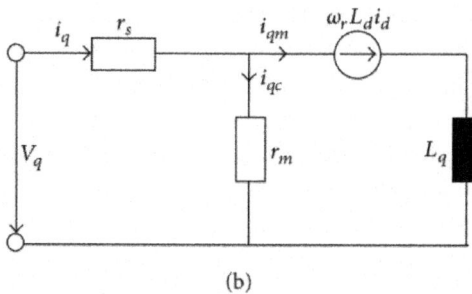

(b)

Figure 3: Equivalent circuits of SynRM in rotor reference d (a) and q frame (b) [8].

For this reason, the equations are expressed in d-q axes, since d-q system leads to more simplified equations and also to a better understanding of the machine's dynamics.

The corresponding equations in the rotor reference frame are the following [8]:

$$V_d = r_s i_{dm} + L_d \left(\frac{di_{dm}}{dt} \right) - \left[1 + \left(\frac{r_s}{r_m} \right) \right] \omega_r L_q i_{qm},$$

$$V_q = r_s i_{qm} + L_q \left(\frac{di_{qm}}{dt} \right) + \left[1 + \left(\frac{r_s}{r_m} \right) \right] \omega_r L_d i_{dm}, \tag{1}$$

where i_{dm} and i_{qm} are given by

$$i_{dm} = i_d + i_{dc} = i_d + \left(\frac{\omega_r L_q i_{qm}}{r_m} \right),$$

$$i_{qm} = i_q - i_{qc} = i_q - \left(\frac{\omega_r L_d i_{dm}}{r_m} \right), \tag{2}$$

where V_d, V_q are the voltages in direct (d) and quadrature (q) axes, i_d, i_q are the currents in d and q axes, w_r is the rotational speed of the frame, L_d and L_q are the inductances in d and q axes, and r_m is an equivalent stator iron loss resistance.

The developed torque is given by the following equation:

$$T_e = \left(\frac{3p}{2} \right) \left(L_d - L_q \right) i_{dm} i_{qm}. \tag{3}$$

From the previous analysis of SynRM, we can make some useful conclusions about the role of the inductances L_d and L_q. From (3), we can derive that a high value in the difference L_d-L_q leads to a high maximum torque. Also, is can be proved that the power factor is mainly affected by the L_d/L_q ratio. Subsequently, it is clear that the SynRM should be designed in such a way that the highest possible value of L_d and the lowest L_q are achieved [8, 23–26]. Nonetheless, there are limits for theses values, since structural restrictions have to be met.

Control Strategies for Synchronous Reluctance Motor

The main characteristics that differentiate this type of motor from other types, such as synchronous and induction motors, are the absence of field winding on the rotor and the fact that there is no rotation slip. Many control methods (more or less complicated) have been proposed for this motor along the years [7, 8, 24, 25, 27,28].

In order to have a simple block diagram of a scalar control method for the SynRM, the open-loop V/f control of the rotor speed rotation speed is presented in Figure 4.

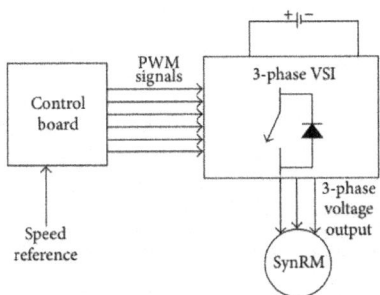

Figure 4: Open-loop control of a SynRM.

The user can set the desired speed and this command is translated to the corresponding voltage signals through a function generator, for example, a microcontroller. These signals are driven to a 3-phase PWM inverter, which is presented in the following chapter. The 3-phase inverter is fed by DC voltage, which is produced through a rectifier topology [24]. Finally, the output voltages are driven to the 3-phase motor and force the rotor to rotate with the desired speed.

VOLTAGE SOURCE INVERTERS

The three-phase VSI topology consists of six power switches (e.g., IGBTs), which are controlled by six corresponding signals (Figure 5).

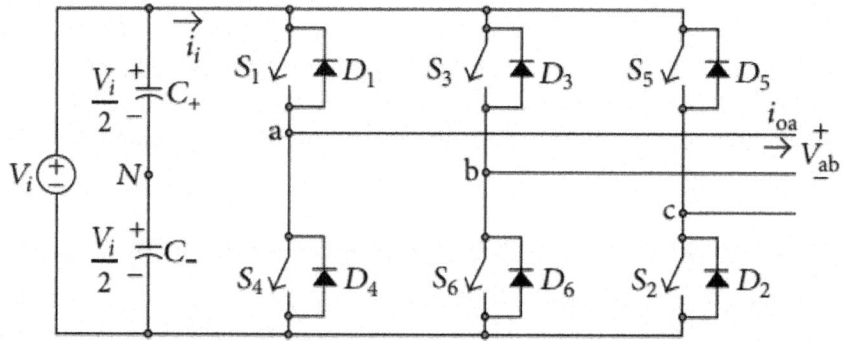

Figure 5: VSI topology.

As it has been already mentioned, PWM techniques are used for the production of the output voltages in a VSI. The main purpose of the VSI is to produce a fundamental voltage harmonic with controlled rms value and frequency.

A properly designed PWM VSI offers the following characteristics:

- high switching frequency, which leads to smaller passive harmonic filters;
- precise regulation of the fundamental harmonic frequency and rms value;
- reduced high-order harmonic content.

Many PWM techniques have been proposed along the years, which in general aimed at the optimization of the harmonic content, better exploitation of the DC voltage source, and reduction of switching losses [29, 30].

Among them, one of the simplest but also the most popular one is the Sinusoidal PWM (SPWM) technique, where three-phase sinusoidar reference signals are compared with a triangular waveform. The corresponding output for each phase is positive when the value of the sine wave is greater than the value of the triangle, and zero in the opposite case. If we denote by m_a the $V_{sine,max}/V_{triangle,max}$ ratio, then the resulting maximum value of the line voltage is equal to $V_{line,rms} = 0.866\, m_a V_i\ (m_a \le 1)$..

Harmonic injection PWM (HIPWM) is a modification of SPWM, where the 3 sine waves contain additionally a small percentage of 3rd and 9th harmonics of the fundamental frequency. This modification leads to better exploitation of the input DC voltage, resulting in higher maximum line, as shown in Figure 6.

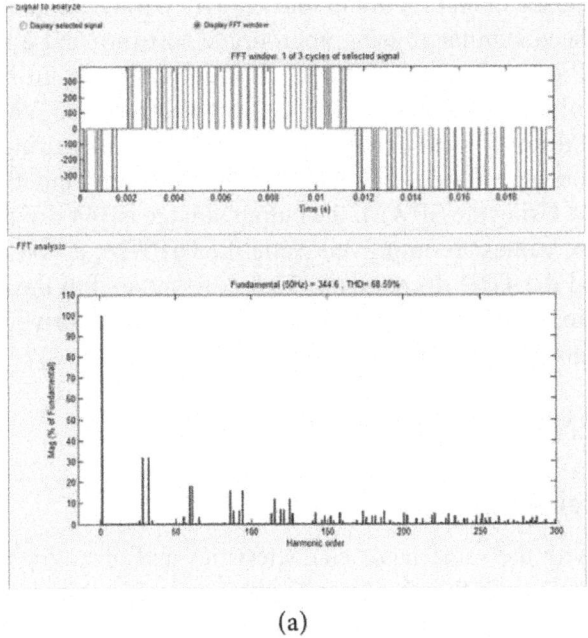

(a)

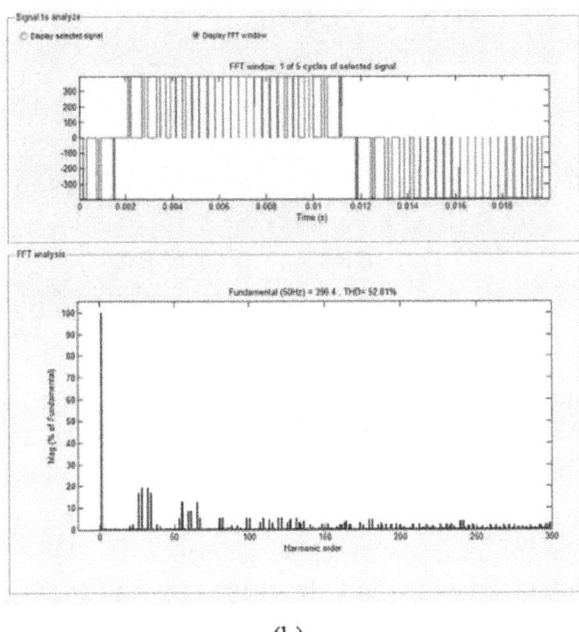

(b)

Figure 6: (a) SPWM and (b) HIPWM voltage signal and FFT analysis.

To evaluate the operation of the selected HIPWM technique, a three-phase inverter has been simulated using appropriate software and a comparison of the classic SPWM and the HIPWM technique was done. In Figure6(a), the output line voltage and its harmonic content using the SPWM technique is presented, while in Figure6(b) the corresponding results using the HIPWM technique are demonstrated. In both cases, the inverter's input DC voltage is equal to 400 V. Using the SPWM, the output voltage is 344.6 V and the THD is 68.59%. These values are improved using the HIPWM, as the output voltage is 396.4 V and the THD decreases to 52.81% compared to the previous case. For this reason, the HIPWM technique has been used to drive the inverter in the experimental investigation that follows.

EXPERIMENTAL SETUP AND RESULTS

Experimental Setup

An inverter with the same basic characteristics and operating with the same HIPWM method as in the simulation has been designed and constructed the main components of the experimental device are shown in Figure 7.

Figure 7: Main components of inverter topology. (1) DC link capacitors, (2) +5 V and +15 V DC supply, (3) control board (Dspic30f4011), (4) Development Board, (5) 3-Phase inverter, and (6) SynRM.

The inverter uses IGBTs as switching elements, since they offer low on-state impedance and adequate switching characteristics [24]. A dsPIC30f4011 microcontroller, programmed in C language, was chosen for the production and control of PWM pulses. The dsPIC peripherals that have been used are analog inputs, which are used in order to read the desired rotation speed (speed reference), digital inputs, one of which is used for ON/OFF function, and a PWM module, which produces the six required control signals for the pulse width modulation of the VSI Bridge. The flow diagram of microcontroller for an open-loop control scheme is given in Figure 8.

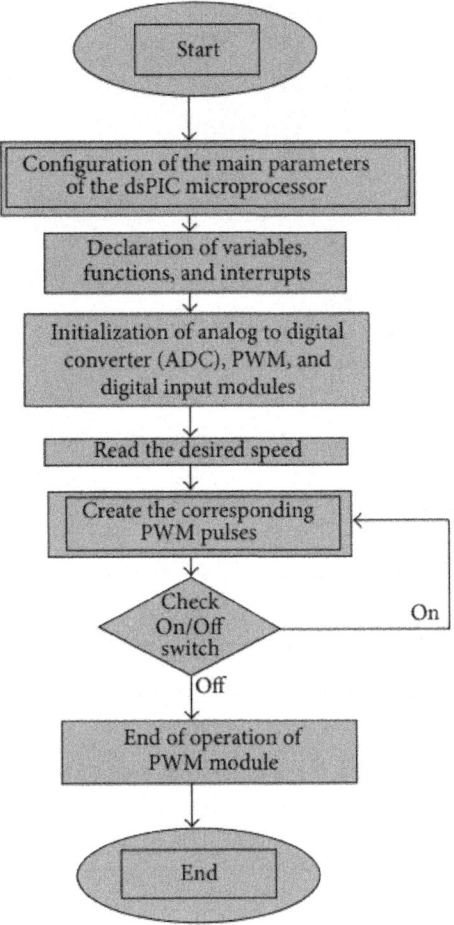

Figure 8: Flow chart of the developed open-loop control technique.

For the experiments, a 4-pole SynRM constructed by Kaiser Motoren is used. Its rotational speed can vary between 150 and 1500 rpm and the nominal power at 50 Hz is 1.26 kW. The rotor of this mass-production motor shows great interest because of its design and construction. It is made from ferromagnetic steel and aluminium alloy, as it is shown in Figure 9. The latter forms a damping cage, which enables the asynchronous starting of the motor by simply connecting the stator to the three-phase network. Additionally, it contributes to the reduction of the oscillations on the rotor. Of course, synchronous starting through an inverter topology is also possible, similarly to SynRM.

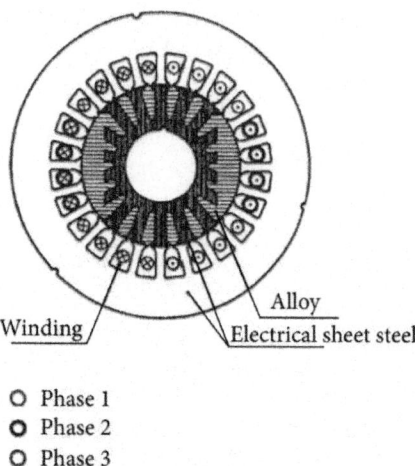

O Phase 1
O Phase 2
O Phase 3

Figure 9: Sketch of the SynRM under investigation.

For evaluating the performance of the system, experimental results have been carried out at no-load condition. This way, a "worst-case scenario" is examined, as oscillations are greater at no-load condition.

Experimental Results

In Figure 10, the pulses of an IGBT in the three-phase VSI which is modulated by HIPWM technique are shown.

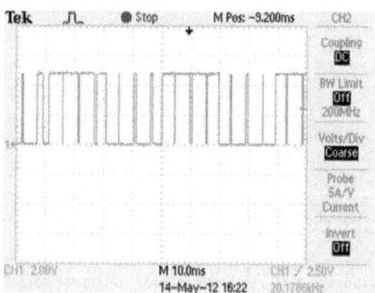

Figure 10: HIPWM gate signal.

As it is mentioned, in the experiments an investigation of the current harmonics has been performed. In Figure 11, the current harmonics are presented, in the case of direct connection to the three-phase network. It is clear that the main higher harmonics are the 5th and the 7th, namely, at 250 Hz and 350 Hz.

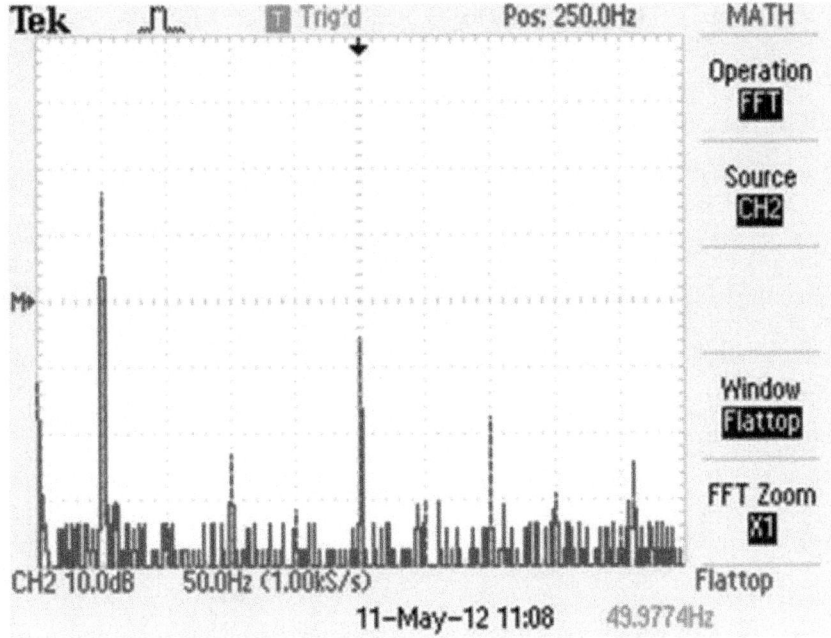

Figure 11: Current harmonics when the motor is fed by the utility grid.

When the topology shown in Figure 7 is used, the rotational speed of the rotor can be adjusted by controlling the corresponding analog input signal of the microprocessor. For the experiments shown, a constant V/f control method was used. As far as the PWM method is concerned, HIPWM was used due to its features which were mentioned before.

With the reference speed set at 50 Hz, the resulting harmonic content of the current is shown in Figure 12. It is clear that, in this case, the main higher harmonics are the 5th and the 7th too, while in general all the other harmonics have higher amplitude compared to the direct connection to the three-phase network. Subsequently, the 5th and 7th harmonics are characteristic frequency components of the stator current, induced by the air gap MMF due to the stator winding distribution. With the reference speed set on 25 Hz, the resulting current waveform and harmonic content are shown in Figures 13 and 14. The harmonics are mainly caused by the rotor field anisotropy due to the construction characteristics of the rotor as it can be seen in Figure 9. Additionally, the torque ripple that is characteristic for this motor (according to the manufacturer) contributes to the rise of the energy contained in higher-order harmonics.

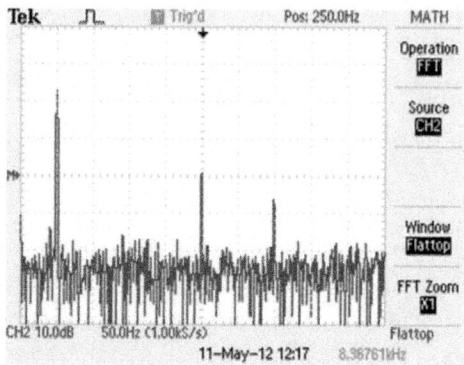

Figure 12: Current harmonics at 50 Hz (fed by a VSI).

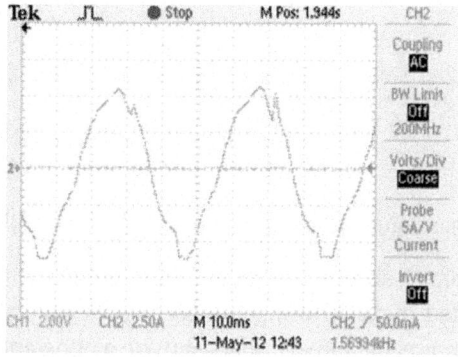

Figure 13: Current Waveform at 25 Hz (actual speed 746 rpm).

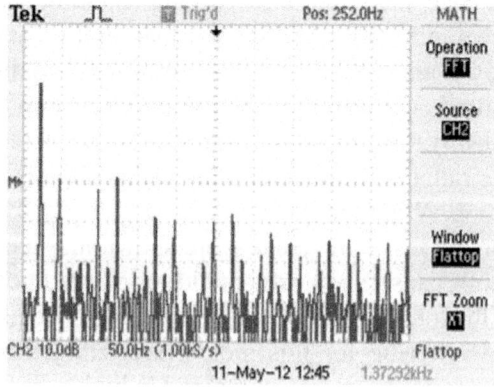

Figure 14: Harmonic content of current at 25 Hz (fed by the VSI).

CONCLUSIONS

A study on the synchronous reluctance motor has been made in this paper. The experimental results have verified the smooth and precise performance of the motor at 50 Hz. Furthermore, it has been observed that the energy of the higher-frequency current harmonics at low speeds, such as 25 Hz, becomes noticeable. To improve performance, a more sophisticated control method for mass-production SynRMs has to be sought. To this direction, the adoption of control methods already available for other synchronous motors has to be considered. In any case, the possibility of using SynRM in all-electric ship applications is subject to further study, since the characteristics of this type of motor show a great promise for the future.

REFERENCES

1. L. Casson, Travel in the Ancient World, The Johns Hopkins University Press, 1994.

2. B. Wagner, All-Electric Ship Could Begin to Take Shape By 2012, NDIAs Business and Technology Magazine, 2007.

3. E. D. Mitronikas and E. M. Tatakis, "Migrating the experience of industrial systems to electric ships: propulsion motors and fault detection," in Proceedings of the 1st MARINELIVE International Workshop on Electric Machines and Power Converters, Athens, Greece, January 2012.

4. C. G. Hodge, D. J. Mattick, and D. J., "The Electric Warhip," Trans IMarE, vol. 108, Part 2, pp. 109–125, 1996.

5. W. J. Kruijt, An Integrated Approach to an All-Electric Cruise Vessel, Business Briefing, Global Cruise, 2004.

6. T. J. McCoy, "Trends in ship electric propulsion," in Proceedings of the IEEE Power Engineering Society Summer Meeting, pp. 343–346, July 2002.

7. H. A. Zarchi, G. R. A. Markadeh, and J. Soltani, "Direct torque and flux regulation of synchronous reluctance motor drives based on input-output feedback linearization," Energy Conversion and Management, vol. 51, no. 1, pp. 71–80, 2010.

8. J. B. Im, W. Kim, K. Kim, C. S. Jin, J. H. Choi, and J. Lee, "Inductance calculation method of synchronous reluctance motor including iron loss and cross magnetic saturation," IEEE Transactions on Magnetics, vol. 45, no. 6, pp. 2803–2806, 2009.

9. T. J. E. Miller, Electronic Control of Switched Reluctance Machines, Newnes Publication, 2001.

10. R. Krishnan, Switched Reluctance Motor Drives, CRC Press, 2001.

11. T. A. Lipo, "Synchronous reluctance machines. A viable alternative for ac drives?" Electric Machines and Power Systems, vol. 19, no. 6, pp. 659–671, 1991.

12. A. Boglietti and M. Pastorelli, "Induction and synchronous reluctance motors comparison," inProceedings of the 34th Annual Conference of the IEEE Industrial Electronics Society (IECON '08), pp. 2041–2044, November 2008.

13. A. A. Arkadan, N. Al-Aawar, and A. A. Hanbali, "Design optimization of SynRM drives for HEV power train applications," in Proceedings of the IEEE International Electric Machines and Drives Conference (IEMDC '07), pp. 810–814, May 2007.

14. C. C. Chan, "The state of the art of electric, hybrid, and fuel cell vehicles," Proceedings of the IEEE, vol. 95, no. 4, pp. 704–718, 2007.

15. N. Schofield, S. A. Long, D. Howe, and M. McClelland, "Design of a switched reluctance machine for extended speed operation," IEEE Transactions on Industry Applications, vol. 45, no. 1, pp. 116–122, 2009. ·

16. V. M. Moreno and A. Pigazo, "Future trends in electric propulsion systems for commercial vessels,"Journal of Maritime Research, vol. 4, no. 2, pp. 81–100, 2007.

17. C. G. Hodge and D. J. Mattick, "The Electric Warship II," in Trans IMarE, vol. 109 of Part 2, The Institute of Marine Engineers, 1997.

18. C. G. Hodge and D. J. Mattick, "The Electric Warship III," in Trans IMarE, vol. 110 of Part 2, The Institute of Marine Engineers, 1998.

19. C. G. Hodge and D. J. Mattick, "The Electric Warship IV," Trans IMarE, vol. 111, Part 1, pp. 25–30, 1999.

20. C. G. Hodge and D. J. Mattick, "The Electric Warship V," in Trans IMarE, vol. 112 of Part 2, The Institute of Marine Engineers, 2000.

21. C. G. Hodge and D. J. Mattick, "The Electric Warship VI," in Trans IMarE, vol. 113 of Part 2, The Institute of Marine Engineers, 2001.

22. J. M. Prousalidis, N. D. Hatziargyriou, and B. C. Papadias, "On studying ship electric propulsion motor driving schemes," in International Conference on Power Systems Transients (IPST '01), Paper 82, pp. 24–28, Rio de Janeiro, Brazil, June 2001.

23. H. A. Toliyat and G. B. Kliman, Handbook of Electric Motors, CRC Press, 2nd edition, 2004.

24. M. H. Rashid, Power Electronics Handbook, Academic Press, 2nd edition, 2001.

25. I. Boldea, Z. X. Fu, and S. A. Nasar, "Torque vector control (TVC) of axially-laminated anisotropic (ALA) rotor reluctance synchronous motors," Electric Machines and Power Systems, vol. 19, no. 4, pp. 533–554, 1991.

26. K. B. Bose, Modern Power Electronics and AC Drives, Prentice Hall, 2001.

27. R. Morales-Caporal and M. Pacas, "A predictive torque control for the synchronous reluctance machine taking into account the magnetic cross saturation," IEEE Transactions on Industrial Electronics, vol. 54, no. 2, pp. 1161–1167, 2007.

28. T. Matsuo, A. El-Antably, and T. A. Lipo, "A new control strategy for optimum-efficiency operation of a synchronous reluctance motor," IEEE Transactions on Industry Applications, vol. 33, no. 5, pp. 1146–1153, 1997.

29. M. A. Boost and P. D. Ziogas, "State-of-the-art carrier PWM techniques: a critical evaluation," IEEE Transactions on Industry Applications, vol. 24, no. 2, pp. 271–280, 1988.

30. S. Jeevananthan, R. Nandhakumar, and P. Dananjayan, "Inverted sine carrier for fundamental fortification in PWM inverters and FPGA based implementations," Serbian Journal of Electrical Engineering, vol. 4, no. 2, pp. 171–187, 2007.

Chapter 3

SINGLE GLUCOSE BIOFUEL CELLS IMPLANTED IN RATS POWER ELECTRONIC DEVICES

A. Zebda[1,2], S. Cosnier[1] , J.-P. Alcaraz[2] , M. Holzinger[1] , A. Le Goff[1], C. Gondran[1], F. Boucher[2], F. Giroud[1], K. Gorgy[1], H. Lamraoui[3] & P. Cinquin[2]

[1]Univ Grenoble 1, CNRS, De´partement de Chimie Moleculaire, UMR-5250, ICMG FR-2607, BP-53, 38041 Grenoble Cedex 9, France

[2]UJF-Grenoble 1/CNRS/TIMC-IMAG UMR 5525, Grenoble, F-38041, France

[3]UroMems - 46 av. Fe´lix Viallet - 38031 GRENOBLE cedex, France.

ABSTRACT

We describe the first implanted glucose biofuel cell (GBFC) that is capable of generating sufficient power from a mammal's body fluids to act as the sole power source for electronic devices. This GBFC is based on carbon nanotube/enzyme electrodes, which utilize glucose oxidase for glucose oxidation and laccase for dioxygen reduction. The GBFC, implanted in the abdominal cavity of a rat, produces an average open-circuit voltage of 0.57 V. This implanted GBFC delivered a power output of 38.7μW, which corresponded to a power density of 193.5 $\mu W\ cm^{-2}$ and a volumetric power of $161\mu W\ mL^{-1}$. We demonstrate that one single implanted enzymatic GBFC can power a light-emitting diode (LED), or a digital thermometer. In addition, no signs of rejection or inflammation were observed after 110 days implantation in the rat.

INTRODUCTION

Since the first successful cardiac pacemaker was implanted in 1960, a variety of implantable battery-powered devices has been developed for various indications, ranging from neurological disorders to hearing loss. The development of lithium batteries in the late 1960s led to better and smaller devices, which showed multiyear longevity and high reliability[1]. Although such batteries continue to be considered as the first choice to power electronic medical implants, there are numerous efforts to develop alternative power-supply systems that are capable to operate independently over prolonged periods of time without the need for external recharging or refuelling[2,3,4]. Several alternatives have been explored in order to power implanted devices with energy from sources in the patient's body. However, systems that take advantage of the Seebeck thermoelectric effect, vibrations or body movements to generate power for an implanted device are limited because these techniques are dependent on the non-continuous nature of vibrations or temperature differences within the human body. GBFCs represent a more promising alternative because they are theoretically able to operate indefinitely due to the ubiquity of glucose and oxygen in the extra-cellular body fluid at constant levels of $5 \times 10^{-3} molL^{-1}$ and $45 \times 10^{-6} molL^{-1}$, respectively[5,6,7,8].

The production of electric power out of body fluids of animals, using glucose as fuel, was first envisioned in the 1970's. In their review, Kerzenmacher et al. mentioned implanted abiotic glucose fuel cells using noble metals as catalysts3. However, the low specificity of the catalysts and the low power output density of these implanted devices precluded further developments.

Following recent developments in nano- and biotechnology, state-of-the-art biofuel cells guarantee high specificity to the fuel, along with satisfactory power densities. These milestones have given rise to a steady growing interest in this research field[9].

Biofuel cells often employ enzymes to catalyze chemical reactions, thereby replacing traditional catalysts present in conventional fuel cells[10,11,12]. These systems generate electricity under mild conditions through the oxidation of renewable energy sources[13]. The advantages of biocatalysts are reactant selectivity, activity under physiological conditions, and facile manufacturability[14].

With the aim of developing implantable power sources in the human body, Katz *and co-workers* demonstrated that a GBFC can produce electricity from a snail[15] and reported more recently two "Cyborg" lobsters connected in-series to power a watch[16]. Rasmussen *et al*[17]placed a GBFC in an insect (Blaberus discoidalis, a cockroach species). Szczupak *et al*[18] implanted GBFCs in clams and connected three of them in series. With this setup, a capacitor could be charged, allowing an electrical motor to rotate. Although these experiments were not performed with mammals, these studies demonstrate that GBFCs can produce electricity out of living organisms. It appears that an attractive mode of operation for biofuel cells consists in the energy accumulation through capacitors for the intermittent activation of model devices. This mode of

GBFC function may be applied to the activation of some sensor devices for medical monitoring.

Concerning the human implantable application, we recently reported a GBFC that is able to generate electric power inside a rat from glucose and oxygen contained in its body fluids[19]. This work was the first demonstration of an implanted GBFC delivering electrical power inside a living organism and, in particular, inside a mammal. However, both, the open-circuit voltage and power density were far below the levels required to supply implanted biomedical devices.

Our recent improvements in GBFC concepts in terms of carbon nanotube compression and direct electron transfer led to high open-circuit voltage (OCV), high power output, and stabilities over weeks[20]. Taking advantage of this improved performance, we report here an original design of a GBFC, based on carbon nanotube-matrix bioelectrodes, and its successful implantation in a rat. One single implanted GBFC device of 0.24 mL volume (2.4 mL for the whole implant) produced the power required to operate, using a specially designed electronic circuit to charge a capacitor, two types of electronic devices: a LED and a digital thermometer.

RESULTS

The bioelectrodes were formed by compression of a CNT/enzyme mixture to pellets and wired using a carbon paste[20]. Such electrodes were wrapped in a dialysis membrane and placed in a perforated silicone tube (Figure 1A), protected by a silicon layer (Figure 1B), packed in dialysis bag, and sutured inside a Dacron® sleeve (Figure 1C). The GBFC was surgically implanted in the abdominal cavity of a rat where the wires were tunnelled up to the head. These wires were then soldered to a female connector and fixed on the skull (Figure 1D and E). Details about the GBFC production and implantation can be found in the methods section.

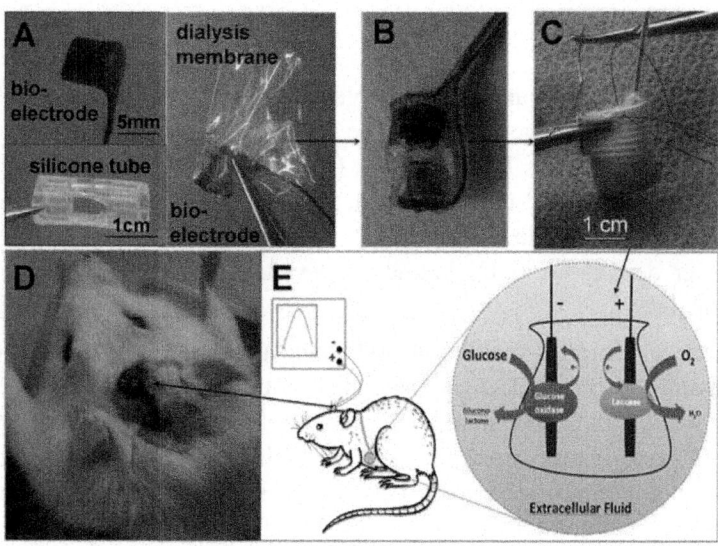

Figure 1: From bioelectrodes to a biocompatible biofuel cell, implanted in the abdominal cavity of a rat.

(A) Image of the components forming the biofuel cell: bioelectrode, perforated silicone tube, and bioelectrode wrapped in a dialysis membrane. (B) Image of our GBFC setup including both, bioanode and biocathode inserted in a silicone cylinder and sealed with silicone. (C) Photograph of the GBFC sutured in a Dacron® bag before implantation. (D) Electrical connection of the implanted GBFC in a Wistar rat; the output wires are fixed to the rat's skull. (E) Schematic description of the enzyme reactions producing electricity, their electrical connection and the GBFC location inside the rat.

The performance of four biofuel cells implanted in four different rats was characterized electrochemically. Figure 2A shows the evolution of the OCV values of four implanted biofuel cells. The measured maximum OCVs, obtained after 6–8 days, were in the range 510–660 mV (Figure 2B) and reflect the difference between the redox potential of laccase at the biocathode and the redox potential of GOx at the bioanode. Figure 2C represents the variation of the current as function of the biofuel cell voltage. For applied current between 150 and 700 $\mu A\ cm^{-2}$ the average voltage measured after 300 s of chronopotentiometry varies from 535 mV to 220 mV. Moreover, these GBFCs can continuously deliver electricity over 10 minutes at an applied discharge current of 150 $\mu A\ cm^{-2}$, producing 9.3×10^{-3} J. It should be noted that the voltage decreased slowly from 0.55 to 0.48 V, as it is displayed in the discharge curve of the implanted GBFC of rat 2 (Figure 2D).

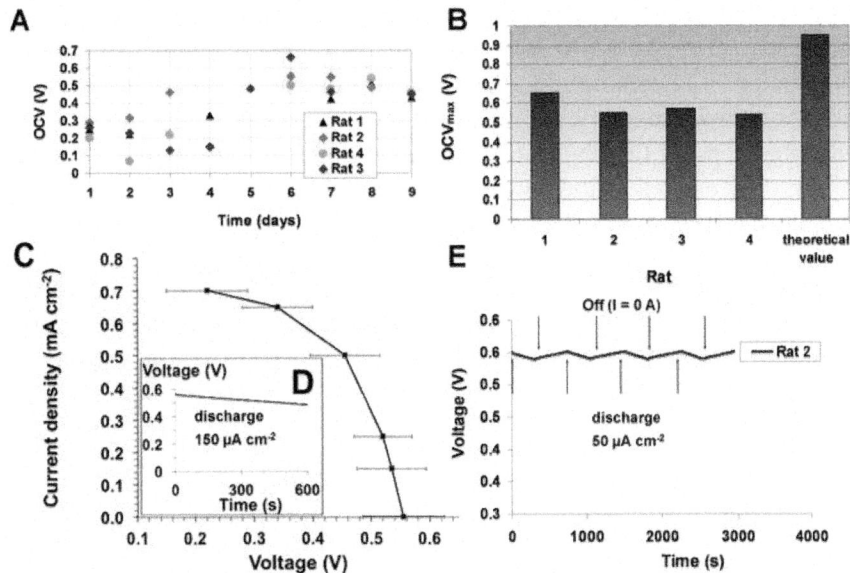

Figure 2: Electrochemical characteristics of the implanted biofuel cell. (A) Evolution of OCV over time for 4 implanted GBFCs (B) Maximum OCVs recorded after 6–8 days; (C) Current versus average voltage. Potential values measured after 300 s of constant current discharges in the range 150–700 μA cm^{-2}. (D) Voltage dependence over time under continuous discharge of 150 μA cm^{-2} (E) Evolution of the GBFC voltage during 4 discharge and stabilization cycles, using each time 50 μA cm^{-2} discharge current for 5 minutes.

To examine more accurately the GBFC voltage evolution, the latter was continuously recorded for an intermittent use of GBFC consisting of 5 minutes discharge at 50 μA cm^{-2} followed by 7 minutes recovery at zero current. For four successive cycles, it appears that, during the discharge period, the GBFC loses less than 20 mV of its initial voltage. Furthermore, the voltage increased when the discharge was stopped, reaching its initial value after 7 minutes (Figure 2E).

These results clearly demonstrate that the implanted GBFCs are able to recover their equilibrium state after the discharge period. The time necessary to reach the initial equilibrium state is most likely due to the consumption of glucose and oxygen at the vicinity or inside the porous bioelectrodes during the GBFC discharge. These successive on-off discharge cycles indicate that this GBFC is able to deliver 50 μA cm^{-2}during a total of 25 minutes of discharge per hour without altering its performance.

With the aim to illustrate the potential of implanted GBFC for powering real electronic devices, the implanted GBFC was connected to a boost converter (step-up converter) and applied to power a light-emitting diode (LED) that consumes 1.31×10^{-3} J (details in the method section).Figure 3A shows the illuminated LED connected to the GBFC via this boost converter. The related video (see supplementary informations) shows 5 successive flashes of the LED using this step-up. Taking into account the average time of energy accumulation for 5 flashes as well as the power efficiency of the boost converter (75%), the implanted GBFC supplied an average power of 38.7 μW or 161 μW mL^{-1}.

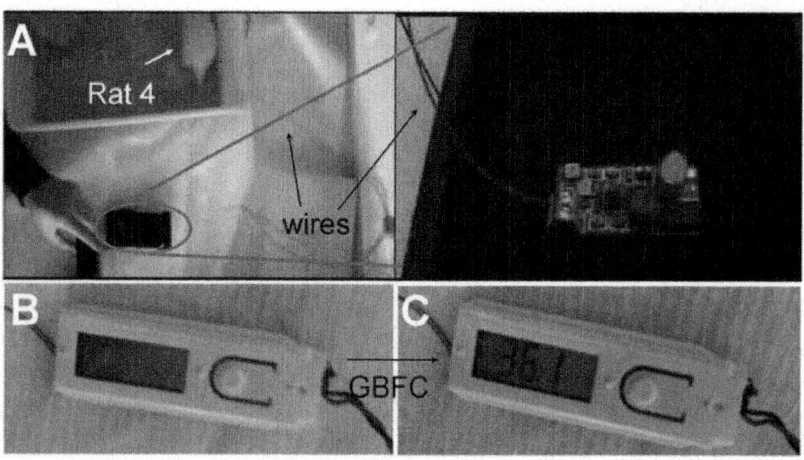

Figure 3: A rat lights a LED and a digital thermometer (A) Image of the LED flashes after its connection to the implanted GBFC (see video in SI), Image of a digital thermometer (B) before and (C) after connection to the implanted GBFC.

In the same way, the implanted GBFC, combined with adapted electric circuitry, was used to power an electric thermometer that requires about 75 μW. Figure 3B and 3C shows the digital display of the powered and non-powered electric thermometer. It appears that the temperature is measured and displayed on an LCD screen when the device is connected to the GBFC. These results clearly show that the implanted GBFC can be considered as a real power source that is able to supply power to medical electronic devices.

Autopsies performed on the rats after 10 to 17 days show that the external side of the implant starts to be covered by a thin layer of vascularized tissues (See for instance rat 4, Figure 4A). After 110 days implantation in rat 1, the implant is surrounded by thick adherent adipose tissue richly vascularized (Figure 4B). The Dacron® bag, conventionally used for implants, ensures excellent biocompatibility for the GBFC. To investigate whether the implanted GBFC can modify the physical fitness of the rat or its eating habits, we monitored daily the weight of the rat and the food consumed during more than 3 months. These observations were monitored with two rats: one rat implanted with a functional GBFC (rat 1) and another with a pseudo-GBFC containing BSA instead of enzymes (control rat). After a small weight loss due to the surgical intervention, the rat>s weight increases gradually in a normal manner (Figure 4C) with an average food intake of 26 ± 3 g per day (Figure 4D). Nevertheless, it should be noted that this behavior does not reflect the biocompatibility of the electronic device in operation for three months due to the loss of connection with the biofuel cell after 9 days.

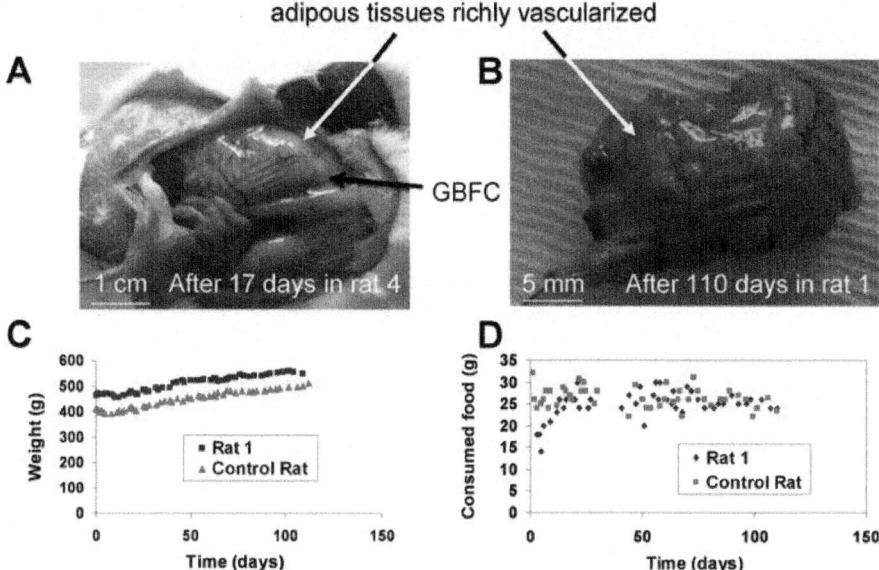

Figure 4: Biocompatibility of the GBFC. Recovered GBFCs after (A) 17 days and (B) 110 days inside a rat. Both implants are surrounded by an adherent adipous tissue richly vascularized. Evolution of (C) the weight and (D) the food intake of the rat with a GBFC (rat 1) and of the control rat for 110 days.

DISCUSSION

The single implanted GBFC that we describe produces the highest output characteristics obtained for a GBFC inside the body of a mammal. One of the principal challenges for the development of an implantable electrochemical energy converter is the biocompatibility, which we achieved by enclosing the GBFC in a Dacron® bag. Moreover, our use of porous dialysis membranes inside the Dacron® bag not only prevents the external leakage of bioelectrode components (CNTs and enzymes) from the GBFC, but also prevents inward diffusion from the body fluids of biological macromolecules which can inhibit the activity of the enzymes.

The GBFC reported here produced a greater power output compared to our first report of an implanted GBFC operating with redox mediators19. In the current GBFC, the use of direct electron transfer between enzymes and electrodes led to a 20-fold increase in power density and an average OCV of 0.57 V obtained with a single GBFC. However, the *in vivo* performance was lower than that observed during the *in vitro*experiment20. This was probably due to the dioxygen concentration in extracellular fluids (which is below 4.5 $\times$ 10^{-5} mol L^{-1} at the venous level) being about four times lower than in the air-saturated buffer solutions that were used for the *in vitro* experiments. All power consuming electrochemical tests of the implanted biofuel cells were tested 6–8 days after surgery and the maximum OCV was measured daily until the power loss of the GBFC. After autopsy of the rats and recovery of the implant, the observed power loss was in most cases due to the dislocation of the wires from the bioelectrodes and sometimes occurred because the rat dislodged the external connector used for the GBFC measurements. In order not to harm the rat, the electrical wires used to connect the cell to the external connector were very flexible and very thin. As a consequence, after about 9 days, we were faced with mechanical breakage of free wires or wire breakage and hence disconnection at the level of the bioelectrode. It is important to note that, unlike all previous experiences made by other research groups with insects, molluscs or lobsters, the animal is not immobilized or anesthetized but remains always freely moving. A clear statement about the lifetime of our biofuel cells is therefore not possible until those engineering issues are resolved.

The obtained power density of our implanted GBFC represents a promising solution to several issues for electronic medical devices. Compared to our previous results19, the volumetric power characteristics represent an 7-fold increase in performance (161 μW mL^{-1} vs 24.4 μW mL^{-1}) for a single implanted GBFC inside the body of a mammal, with an excellent biocompatibility ensured by the Dacron® bag. The GBFC that we report produces significant levels of energy at a single location and hence could be utilized as the power source for implanted sensor devices dedicated to medical monitoring. Further optimization of our GBFC could be expected to provide opportunities for other medical applications such as Multiple-lead Cardiac Resynchronisation Therapy (promising for some types of Heart Failures), or peripheral nerve stimulations (e.g. for pain control). In the case of peripheral nerve stimulation several small GBFCs could be used instead of the currently implanted bundle of wires from a single sealed battery.

METHODS

Ethics statements

The care of the rats was approved by the European Communities Council Directive Animal Care and Use Committee. The experiments were performed in accordance to their guiding principles (European Communities Council Directive L358-86/609/EEC). All protocols involving living animals were performed under license from the French Ministry of Agriculture (License number 38018 and 381141). The Committee on the Ethics of Animal Experiments of the Grenoble University (ComEth) approved the protocol under the number103_LER-PRETA-JPA-01).

Chemicals

Commercial thin Multi-walled Carbon Nanotubes (9.5 nm diameter, purity > 95%) obtained from Nanocyl were used as received without any purification step. Glucose oxidase (GOx) from *Aspergillus niger* (100 U mg^{-1} solid), laccase from *Trametes versicolor* (20 U mg^{-1} solid) were purchased from Sigma–Aldrich and used without further purification. Cellulose membranes were purchased from Spectrumlabs: Spectra/Por® Dialysis membrane, MWCO 6–8000 g mol^{-1}, flat width 32 mm, diameter 20.4 mm, vol/length 3.3 mL/cm.

Instrumentation

The electrochemical characterization and the biofuel cell tests were performed with an Autolab potentiostat 100 (Eco Chemie, Utrecht, The Netherlands).

Procedures to Fabricate the Bioelectrodes and the GBFC

The enzymes are mechanically confined into a CNT matrix by compression of a CNT/enzyme mixture as reported20, affording direct electrical wiring between the redox active center and the electrode. Furthermore, the CNT matrix ensures high conductivity and high porosity, necessary for the diffusion of substrates. The bioelectrodes used for implantation are pellets with 6 mm thickness and 5 mm in diameter. Although only laccase is used at the cathode, GOx is combined with catalase at the anode. Catalase catalyzes the decomposition of hydrogen peroxide, produced by unwired GOx and hence prevents enzyme deactivation and serves for local oxygen depletion21. The whole GBFC was then formed by connecting the bioanode and biocathode.

Miniaturized insulated wires (UBA3219, industrifil) are connected to the CNT pellets via a conductive ink (Electrodag 423SS) (see manuscript:Fig.

1A bottom). Each pellet was then wrapped in a cellulose acetate membrane (see manuscript: Fig. 1A middle). These electrodes were then inserted into a perforated silicone sleeve (inner diameter: 5 mm) (see manuscript: Fig. 1A top) and enclosed by a silicone ring. Before use, the dialysis membrane, solutions and catheters were sterilized using an autoclave. The wires were then introduced into a biocompatible silicone tubing (721048, Harvard Apparatus; 1.9 mm outside diameter) that were then filled to a length of 3 cm with medical grade silicone. Silicone was also used to cover the non-biocatalytic parts of the silicone mould (see manuscript: Fig. 1.B). Thus, no wound or trauma can be imputed to the electrical circuitry and the wires were protected from short circuits.

The bagging of the implanted cell was carried out under a laminar flow hood under sterile conditions. The silicone sleeve containing the electrodes is inserted in a 16 mm flat width dialysis bag with a 100–500 Daltons MWCO (131054 Spectrumlabs). The bag is then filled with approximately 1 mL sterile Ringer solution and closed while avoiding air bubbles. The biocompatibility of the resulting device in the rat is due to an autoclaved Dacron® sleeve, wrapped around the dialysis bag, cut to the right volume, and then sutured with surgical filament (see manuscript: Fig. 1C).

Surgical Implantation of the GBFC inside a Rat

Male Wistar rats weighing 300–560 g were anesthetized with isoflurane under inhalational conditions. A median laparotomy was performed to insert the implant into the retroperitoneal space in left lateral position. The catheters containing the wires of the GBFC are subcutaneously tunnelled from the abdomen up to the head of the rat (see manuscript:Fig. 1D and 1E). The wires were then soldered to a female micro-connector (BL3.36Z fischer electronik) which was insulated and fixed to the skull by acrylic cement22. The GBFC was implanted in the retroperitoneal space (Fig. 1E) of the rat because the composition of the extracellular fluid in terms of glucose and oxygen is the same as in blood. The muscular abdominal wall and the skin were finally sutured separately and the animals allowed to recover from anaesthesia. After surgery, the animals received a single injection of an analgesic (Rimadyl, 5 mg kg^{-1}, i.m.).

After implantation, the rats were left to recover 24 hours. Each day, the rats were connected to a potentiostat to measure the open circuit voltage of the implanted GBFC. The rats were not immediately sacrificed after the performance studies in order to evaluate the biocompatibility of the implants. As required, the rats were euthanized under anaesthesia (sodium pentobarbital 50 mg kg^{-1}, i.p.) by intra-cardiac injection of sodium pentobarbital (100 mg).

Power Management of the Glucose Biofuel Cell

Two different electronic devices have been designed to demonstrate the power management of the implanted biofuel cell. One is for powering a common medical digital thermometer (power consumption: 50 μA at 1.5 V) and another to light a LED (4.1 mA at 2.9 V). Both electronic designs are based on a low input voltage boost converter (BQ25504, Texas instruments, Dallas, Texas, USA). The global efficiency of this circuit is well-suited to these devices because it is optimized to operate within the GBFC›s characteristics (input voltage: 0.3 V to 0.6 V, input current: 10 to 100 μA, output voltage: up to 3 V). For instance, the power efficiency of the voltage boost converter was around 75% for the LED demonstrator (average input voltage: 0.5 V for an average current of 70 μA and average output voltage of 2.9 V for an average current of 4.1 mA). The principle of the power management is the same for both demonstrators (thermometer and LED), whereby a capacitor is charged by the GBFC driving the boost voltage converter. When the capacitor voltage reaches a predetermined value, the stored power is released until the capacitor voltage decreases down to a determined value. The cycle is repeated as long as the device is charged by the biofuel cell. In the case of the demonstrator to power-on the thermometer, a capacitor value of 220 μF was chosen in order to keep the temperature displayed on an LCD screen for 10 s after a capacitor charge cycle of one minute. Regarding the LED demonstrator, the diode flashed for about 88 ms after 28 s, 52 s, 81 s, 115 s, and 169 s.

ACKNOWLEDGEMENTS

The authors would like to thank the Laboratory of Bioclinical Radiopharmaceutics (LRB) for providing rats. The authors also thank the Interdisciplinary Energy program of CNRS PR10-1-1, the ANR Emergence-2010 EMMA-043-02, and the ANR Investissements d'avenir - Nanobiotechnologies 10-IANN-0-02 programmes for financial support. This work was supported by French state funds managed by the ANR within the Investissements d'Avenir programme (Labex CAMI) under reference ANR-11-LABX-0004. Arielle Le Pellec is greatly acknowledged for technical assistance.

REFERENCES

1.   Holmes, C. F. Electrochemical Power Sources and the Treatment of Human Illness. *J. Electrochem. Soc. Interface* 12 (3 (Fall)), 26 (2003).

2.   Roundy, S. On the Effectiveness of Vibration-based Energy Harvesting. *J. Intell. Mater. Syst. Struct.* 16 (10), 809 (2005).

3.   Kerzenmacher, S., Ducrée, J., Zengerle, R. & von Stetten, F. Energy

harvesting by implantable abiotically catalyzed glucose fuel cells.*J. Power Sources* 182 (1), 1 (2008).

4. Görge, G., Kirstein, M. & Erbel, R. Microgenerators for Energy Autarkic Pacemakers and Defibrillators: Fact or Fiction? *Herz* 26(1), 64 (2001).

5. Mano, N., Mao, F. & Heller, A. A Miniature Biofuel Cell Operating in A Physiological Buffer. *J. Am. Chem. Soc.* 124 (44), 12962 (2002).

6. Mano, N., Mao, F. & Heller, A. Characteristics of a Miniature Compartment-less Glucose/O_2 Biofuel Cell and Its Operation in a Living Plant. *J. Am. Chem. Soc.* 125 (21), 6588 (2003).

7. Chen, T. *et al.* A Miniature Biofuel Cell. *J. Am. Chem. Soc.* 123 (35), 8630 (2001).

8. Rapoport, B. I., Kedzierski, J. T. & Sarpeshkar, R. A Glucose Fuel Cell for Implantable Brain-Machine Interfaces. *PLoS ONE* 7 (6), e38436 (2012).

9. Holzinger, M., Le Goff, A. & Cosnier, S. Carbon nanotube/enzyme biofuel cells. *Electrochim. Acta* 82, 179 (2012).

10. Minteer, S. D., Liaw, B. Y. & Cooney, M. J. Enzyme-based biofuel cells. *Curr. Opin. Biotechnol.* 18 (3), 228 (2007).

11. Atanassov, P. *et al.* Enzymatic Biofuel Cells. *The Electrochemical Society Interface* 16 (2), 28 (2007).

12. Sarma, A. K., Vatsyayan, P., Goswami, P. & Minteer, S. D. Recent advances in material science for developing enzyme electrodes.*Biosens. Bioelectron.* 24 (8), 2313 (2009).

13. Barton, S. C., Gallaway, J. & Atanassov, P. Enzymatic Biofuel Cells for Implantable and Microscale Devices. *Chem. Rev.* 104 (10), 4867 (2004).

14. Barton, S. C. in *Handbook of Fuel Cells: Advances in Electrocatalysis, Materials, Diagnostics and Durability, Volumes 5 & 6*, edited by W. Vielstich, H. A. Gasteiger & H. Yokokawa (John Wiley & Sons, Weilheim, 2009), Vol. 5.

15. Halámková, L. *et al.* Implanted Biofuel Cell Operating in a Living Snail. *J. Am. Chem. Soc.* 134 (11), 5040 (2012).

16. MacVittie, K. *et al.* From "Cyborg" Lobsters to a Pacemaker Powered by Implantable Biofuel Cells. *Energy Environ. Sci.* 6 (1), 81 (2013).

17. Rasmussen, M. *et al.* An Implantable Biofuel Cell for a Live Insect.*J. Am. Chem.Soc.* 134 (3), 1458 (2012).

18. Szczupak, A. *et al.* Living battery - biofuel cells operating in vivo in clams. *Energy Environ. Sci.* 5 (10), 8891 (2012).

19. Cinquin, P. *et al.* A Glucose BioFuel Cell Implanted in Rats. *PLoS ONE*

5 (5), e10476 (2010).

20. Zebda, A. *et al.* Mediatorless high-power glucose biofuel cells based on compressed carbon nanotube-enzyme electrodes.*Nature Communications* 2, 370 (2011).

21. Plumeré, N., Henig, J. & Campbell, W. H. Enzyme-Catalyzed O_2 Removal System for Electrochemical Analysis under Ambient Air: Application in an Amperometric Nitrate Biosensor. *Anal. Chem.* 84(5), 2141 (2012).

22. Deransart, C. *et al.* Single-unit Analysis of Substantia Nigra Pars Reticulata Neurons in Freely Behaving Rats with Genetic Absence Epilepsy. *Epilepsia* 44 (12), 1513 (2003).

Chapter 4

TM-TE DECOMPOSITION OF POWER LOSSES IN MULTI-STRANDED LITZ-WIRES USED IN ELECTRONIC DEVICES

C. Carretero[1], J. Acero[1], and R. Alonso[2]

[1]Dep. Ingenier´ıa Electr´onica y Comunicaciones, Universidad de Zaragoza, Mar´ıa de Luna, 1, 50018 Zaragoza, Spain

[2]Dep. F´ısica Aplicada, Universidad de Zaragoza, Pedro Cerbuna, 12, 50009 Zaragoza, Spain

ABSTRACT

Efficiency often constitutes the main goal in the design of a power system because the minimization of power losses in the magnetic components implies better and safer working conditions. The primary source of losses in a magnetic power component is usually associated with the current driven by the wire, which ranges from low to medium frequencies. New power system tendencies involve increasing working frequencies in order to reduce the size of devices, thus reducing costs. However, optimal design procedures involve increasingly complex solutions for improving system performance. For instance, using litz-type multi-stranded wires which have an internal structure to uniformly share the current between electrically equivalent strands, reducing the total power losses in the windings. The power losses in multi-stranded wires are generally classified into conduction losses and proximity losses due to currents induced by a magnetic field external to the strand. Both sources of loss have usually been analyzed independently, assuming certain conditions in order to simplify the derivation of expressions for calculating the correct values. In this paper, a unified analysis is performed given that both power losses are originated by the electromagnetic fields arising from external sources where the wire is immersed applying the decomposition into transversal magnetic (TM) and transversal electric (TE) components. The classical power losses, the so called conduction and proximity losses, can be calculated considering the TM modes under certain conditions. In addition, a new proximity loss contribution emerges from the TE modes under similar conditions.

INTRODUCTION

Many models have been developed over the years to calculate power losses produced in windings in order to optimize magnetic power components, such as transformers and inductors. Early results were obtained from analytical models assuming certain conditions in order to simplify the problem. Power loss expressions considering straight cylindrical wires are provided in [1] for different excitations associated with different kinds of power losses. For instance, conduction power losses are associated with an axial electric field whereas proximity losses are related with an external transversal magnetic field. This book provides analytical solutions for simplified systems based on Bessel functions arising from their rotational symmetry. Reviews of several techniques employed in the estimation of power losses in the wires of transformers and inductors are provided in [2–4] with concise evaluations being made by comparison. The method described in [5] is widely used in the design of foil transformers because the power losses are derived from a simplified 1-D modeling of the electromagnetic field, considerably reducing the geometrical complexity and consequently simplifying the numerical calculation. This paper constitutes the basis of a large number of subsequent works [6, 7], but assuming restrictive approximations.

In more recent years, preferred solutions have involved the calculation of the electromagnetic fields of systems by means of analytical or numerical calculations with the power losses being subsequently derived in a straightforward manner from the analytical expressions. The basic structure of typical wires used in high frequency power applications are packed bundles of many small circular crosssection strands made of a good conductor, for instance, copper or aluminum. Multi-stranded wires can be classified depending on the braid of the construction as twisted multi-stranded wires, with the strands occupying a constant distance to the wire axis [8], and litz-type wire, with strands exhibiting radial and azimuthal transposition [9].

The losses in each strand can be analytically calculated by means of two-dimensional equations obtained for a simplified geometry of a straight homogeneous cylinder immersed in a uniform field. This approach has been followed by many authors arriving at expressions based on Bessel functions which are applied to calculate the losses in litz-wires, as may be seen in [10–16]. It should be noted that the solution can be achieved either by considering physical external field excitations [17, 18], or from a Helmholtz potential point of view [19, 20]. Alternatively, simpler expressions can be obtained neglecting the influence of the strand induced currents over the external fields, inother words, the electromagnetic field is not distorted by the presence of the strand. Equivalently, in the latter case, a low-frequency approach is assumed with respect to the previous methodology, as followed in [21–23].

The expressions given in the preceding papers include the electromagnetic fields obtained either by means of analytical techniques [24, 25], or by using numerical methods, in particular methods based on finite elements [23–31].

Litz-wires generally consist of a large number of small size strands, thus the effect of an individual strand in the electromagnetic behavior of the system can be neglected because the fields are slightly distorted. However, their effects are included by addition of the power loss contribution in the equivalent resistance of the device. Moreover, the electromagnetic field where the strand is immersed can be considered external to the strand due to the sources being outside its cylindrical volume. As a consequence, the well-know decomposition into TM and TE components with respect to an arbitrary axis can be applied, as is given in [32–37], to split the external field along the longitudinal axis of the strand. Each component of the decomposition individually contributes to the total power losses induced in the cylindrical conductor, as appears for the TM modes in [38] and for the TE modes in [39]. In this paper, the classical contributions to power losses, namely proximity and conduction losses, are associated with the zeroth and first order TM modes, whereas the zeroth order TE mode concerns the proximity effect due to an external longitudinal magnetic field.

LOSS MODELING IN CYLINDRICAL STRANDS

The internal structure of a multi-stranded wire consists of $n0$ cylindrical strands of circular cross-section made of a material with high electrical conductivity, usually copper or aluminum. Magnetic materials are not suitable for building wires due to the increase in power losses. Generally, the strands are braided in a litz-wire structure in order to minimize the total power losses. This is because such structure, shown in Figure 1, produces greater equivalence between strands and therefore the strands tend to carry the same current.

The analysis of power losses is performed assuming the following conditions. In the first place, a single strand model is developed in order to simplify the treatment. The strand to be analyzed is a straight cylinder of infinite length immersed in the external field. The effect of the strand curvature is therefore neglected. The analysis can be extended to multi-stranded systems by applying the equivalencebetween strands. Second, the power losses arise from the external electromagnetic field which model the effect of the remainder of the system where the strand is placed.

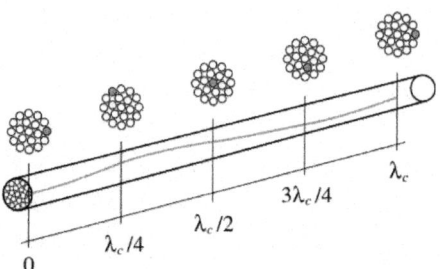

Figure 1: Structure of the litz-wire showing the trajectory of a strand with a characteristic length λ_c.

The strand is therefore assumed to be inside a volume with no field sources, thus the effects of the fields are decoupled from their sources. The decomposition into TM and TE components of the fields is assumed to obtain the complete description of the external field. The longitudinal direction of the strand $\hat{z}$ is chosen to perform this decomposition. Moreover, the magneto-quasistatic approach is assumed to simplify the analysis because the radiation contribution is slight with respect to the diffusion effects of the field inside the conductor [40], or, equivalently, the displacement current J_d in the system is negligible compared to the induced currents J_c in the strand at the working frequencies of the power devices ranging from dc to several MHz. Finally, the power losses are associated with induced current in the strands because the hysteresis effects can be neglected in the media with permeability values close to vacuum permeability μ_0, as occurs for copper or aluminum.

The TM-TE decomposition of the electromagnetic field in the free space where the strand is immersed is performed by means of the longitudinal components $E_z^0(\mathbf{r})$ and $H_z^0(\mathbf{r})$, respectively, which can be expanded in a Taylor series with respect to the reference position $\mathbf{r}_i$, as is shown as follows:

$$E_z^0(\mathbf{r}) = \sum_{n=0}^{\infty} \frac{1}{n!} \left(\mathbf{r}' \cdot \nabla \right)^n E_z^0(\mathbf{r}_i) \quad \text{(TM modes)},$$
(1)

$$H_z^0(\mathbf{r}) = \sum_{i=0}^{\infty} \frac{1}{n!} \left(\mathbf{r}' \cdot \nabla \right)^n H_z^0(\mathbf{r}_i) \quad \text{(TE modes)},$$
(2)

where r' is defined as $\mathbf{r} - \mathbf{r}_i$.

The terms of the expansion applied to the longitudinal electric field $E_z^0(\mathbf{r})$, in (1) are:

$$E_z^0(\mathbf{r}) = E_z^0(\mathbf{r}_i) + \left(\mathbf{r}' \cdot \nabla \right) E_z^0(\mathbf{r}_i) + \ldots,$$
(3)

where the operator ∇ may be divided into transversal ∇_t and longitudinal $\hat{z}\partial_z$ components [42], where the transversal component ∇_t can be expressed in the appropriate coordinate framework, for instance, a rectangular or polar coordinate system. Furthermore, the spatial vector r' is decomposed into the transversal t' and the longitudinal z' vectors. Consequently, we obtain that (3) can be rewritten as:

$$E_z^0(\mathbf{r}) = E_z^0(\mathbf{r}_i) + (\mathbf{t}' \cdot \nabla_t)\, E_z^0(\mathbf{r}_i) + z' \cdot \partial_z E_z^0(\mathbf{r}_i) + \cdots \tag{4}$$

The first term in (4) is a uniform longitudinal electric field $E_z^0(\mathbf{r}_i)\hat{z}$ constituting the main source of the conduction losses in the strands because the conduction current density J_c equal to $\sigma E_z^0(\mathbf{r}_i)\hat{z}$ dissipates energy. The surface integral of J_c in the cross-section area is the current I_0 carried by the strand.

The second term in (4) is an anti-symmetric electric field with respect to the plane at the position r_i with normal directed along the transversal gradient of the electric field. This contribution is associated with a uniform transversal magnetic field $\mathbf{H}_t^0(\mathbf{r})$, which is therefore the source of the classical proximity losses. The relationship between the term $(\mathbf{t}' \cdot \nabla_t)E_z^0(\mathbf{r}_i)\hat{z}$ and $\mathbf{H}_t^0(\mathbf{r})$ can be carried out applying Maxwell's equation:

$$\nabla \times \mathbf{E}(\mathbf{r}, t) = -\partial_t \mathbf{B}(\mathbf{r}, t). \tag{5}$$

In the preceding expression, the electric field E(r) and the magnetic induction field B(r) are equal to $(\mathbf{t}' \cdot \nabla_t)E_z^0(\mathbf{r}_i)\hat{z}$ and $\mu_0 \mathbf{H}_t^0(\mathbf{r})$, respectively. Moreover, the operator ∂_t is replaced by the operator $j\omega$ because the harmonic approach will be implicitly applied, assuming e $^{j\omega t}$ time dependence. As a result, we obtain the following relationship:

$$\nabla \times \left(\hat{z}\, (\mathbf{t}' \cdot \nabla_t)\, E_z(\mathbf{r}_i) \right) = -j\omega\mu_0 \mathbf{H}_t^0(\mathbf{r}). \tag{6}$$

Rearranging the preceding expression, we have:

$$-\hat{z} \times \nabla_t \left((\mathbf{t}' \cdot \nabla_t)\, E_z(\mathbf{r}_i) \right) = -j\omega\mu_0 \mathbf{H}_t^0(\mathbf{r}). \tag{7}$$

Performing the vector product by the unitary vector $\hat{z}$ and afterwards applying the scalar product by the vector $\hat{t}$ in both sides of the equation, the following equivalence is obtained:

$$(\mathbf{t}' \cdot \nabla_t)\, E_z(\mathbf{r}_i) = -j\omega\mu_0 \left(\hat{z} \times \mathbf{H}_t^0(\mathbf{r}) \right) \cdot \mathbf{t}'. \tag{8}$$

Consequently, the second term in (4) can be evaluated in a straightforward manner including the uniform transversal magnetic field $\mathbf{H}_t^0(\mathbf{r})$ in (8).

The last term shown in (4) concerns the longitudinal first order variation of the longitudinal electric field related with the electric charge density in the strand, modifying the current density along the wire, but without associated current densities constituting new power loss sources. A further analysis can be performed considering higher order terms, but the results would be beyond the scope of this paper. In conclusion, the zeroth and first order TM modes originated the classical behavior for the conduction resistance and proximity losses due to the uniform longitudinal electric field $E_z^0(\mathbf{r}_i)\hat{\mathbf{z}}$ and transversal magnetic field $\mathbf{H}_t^0(\mathbf{r})$, respectively.

The complete decomposition of the external electromagnetic field is given by the TM-TE decomposition. The power losses in the strands are typically accounted for considering TM modes only, but TE modes are included in order to achieve a complete description of the power losses. The zeroth term in (2) is a uniform longitudinal magnetic field $H_z^0(\mathbf{r}_i)\hat{\mathbf{z}}$. Higher order terms can be neglected because their contribution to the losses is slight. Note that the zeroth order TE mode constitutes a new proximity loss source.

ANALYTICAL EXPRESSIONS OF POWER LOSSES

The power loss expressions can be obtained starting with the governing equation applied to the corresponding transversal representation of the external field. The cylindrical coordinate system formulation is employed to work out the solution due to the geometrical symmetry of the single strand system.

TM mode fields obey the following scalar equation [38]:

$$\nabla^2 E_z(\mathbf{r}) - j\omega\mu\sigma E_z(\mathbf{r}) = 0, \tag{9}$$

being μ and σ, the magnetic permeability and electric conductivity of the different media, respectively. Moreover, the magnetoquasistatic approach [40] has been applied because the radiation can be neglected with respect to the induced current effects, obtaining an electromagnetic diffusion equation [41]. The solution should be worked out applying the boundary condition for the electric and magnetic field at the surface of the strand and, additionally, the far away fields must converge to the considered external field term. The general solution is based on Bessel and trigonometric functions [42].

Considering the zeroth order term of (4), by solving (9), we obtain [1]:

$$E_z(\mathbf{r}) = \frac{J_0\left((j-1)\rho/\delta\right)}{J_0\left((j-1)r_0/\delta\right)} E_z^0(\mathbf{r}_i) \quad \rho \leq r_0,$$

(10)

And

$$E_z(\mathbf{r}) = E_z^0(\mathbf{r}_i) \quad \rho > r_0,$$

(11)

where δ is the strand penetration depth defined as $\sqrt{2/(\omega\mu\sigma)}$, and ρ is the radial distance with respect to the position r_i located at the center of the strand. Note that the external uniform field is constant in the medium surrounding the strand, but its amplitude diminishes when it penetrates inside the conductive media.

The first order TM term associated with the second term in (4) obeys the following solution [19]:

$$E_z(\mathbf{r}) = 2\frac{1-j}{\sigma\delta}\frac{J_1\left((j-1)\rho/\delta\right)}{J_0\left((j-1)\rho/\delta\right)} \sin\left(\varphi - \varphi_0\right) H_t^0 \quad \rho \leq r_0,$$

(12)

and

$$E_z(\mathbf{r}) = \left[2\frac{1-j}{\sigma\delta}\frac{J_1((j-1)\rho/\delta)}{J_0((j-1)\rho/\delta)}\frac{r_0}{\rho} + j\omega\mu_0\left(\rho - \frac{r_0^2}{\rho}\right)\right]\sin(\varphi - \varphi_0)H_t^0$$

$$\rho > r_0,$$

(13)

where H_t^0 is the magnitude of the transversal magnetic field $\mathbf{H}_t^0(\mathbf{r}_i)$ and $\varphi 0$ is the azimuthal coordinate φ of the vector $\hat{\mathbf{z}} \times \mathbf{H}_t^0(\mathbf{r}_i)$ expressed in the cylindrical reference system. The magnetic field is influenced by the induced currents in the vicinity of the strand.

Both power loss contributions can be associated with resistive loss terms arising from the current densities due to the internal electric field $E_z(\mathbf{r})$ in (10) and (12) for the conduction and proximity losses, respectively. These resistive contributions are analyzed as follows. It should be noted that the reference coordinate r_i has been chosen in such a way as to avoid the contribution of the second term in (4) to the total current carried by the strand. Thus, the conduction resistance can be calculated from the ratio between the voltage per unit of length equal to $E_z^0(\mathbf{r}_i)$ and the current I_0 carried by the strand. Integrating the current density $\sigma E_z^0(\mathbf{r}_i)\hat{\mathbf{z}}$ flowing through the cross-section area, we obtain the current I_0 driven by the cylinder [19]:

$$I_0 = \frac{2\pi r_0 \delta \sigma J_1\left((j-1)r_0/\delta\right)}{(j-1)J_0\left((j-1)r_0/\delta\right)} E_z^0(\mathbf{r}_i).$$

$$(14)$$

The conduction resistance per unit length $R_{cond,\ u.l.}$ is defined as the real part of the ratio between the voltage per unit length, given by the amplitude of $E_z^0(\mathbf{r}_i)$, and the current I_0, hence:

$$R_{cond,\ u.l.} = \frac{1}{\pi r_0^2 \sigma} \Phi_{cond}\left(r_0/\delta\right),$$

$$(15)$$

where the factor $\Phi_{cond(r0/\delta)}$ is defined as:

$$\Phi_{cond}\left(r_0/\delta\right) = \Re\left((j-1)\frac{r_0}{\delta}\frac{J_0\left((j-1)r_0/\delta\right)}{J_1\left((j-1)r_0/\delta\right)}\right),$$

$$(16)$$

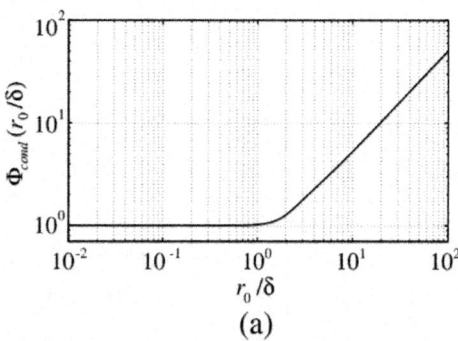

(a)

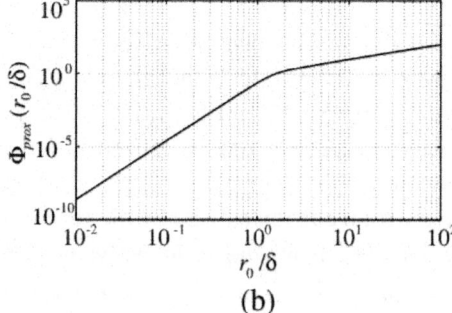

(b)

Figure 2: Representation of conduction and proximity factor depending on the ratio r0/δ between the radius of the strand and the penetration depth. (a) $\Phi_{cond}\left(r_0/\delta\right)$. (b) $\Phi_{prox}^{low}\left(r_0/\delta\right)$.

Note that (15), which is depicted in Figure 2(a), includes the skin effect accounted by the factor $\Phi_{cond(r_0/\delta)}$ depending on the ratio r_0/δ.

The second term in (4) originates the classical proximity power losses in the strand which may be evaluated by integrating the density factor $\frac{1}{2}\mathbf{E}\cdot\mathbf{J}^*$, where E is provided in (12), obtaining the losses per unit length of the strand due to a transversal magnetic field [19]:

$$P_{prox\,t,\,u.l.} = \frac{2\pi}{\delta}\Phi_{prox}(r_0/\delta)\left|H_t^0\right|^2 , \tag{17}$$

where $\left|H_t^0\right|$ is the amplitude of the external magnetic field $\mathbf{H}_t^0(\mathbf{r}_i)$. The factor $\Phi_{prox(r_0/\delta)}$, shown in Figure 2(b), is provided by many authors in a form depending on Kelvin functions, as it appears in [1, 43], but the mathematical equivalent expression given in [27, 29] is preferred because it is more compact, as can be seen as follows:

$$\Phi_{prox}(r_0/\delta) = \Re e\left(j\left(\frac{r_0}{\delta}\right)^2 \frac{J_2\left((j-1)r_0/\delta\right)}{J_0\left((j-1)r_0/\delta\right)}\right), \tag{18}$$

In many cases, the external magnetic field arises from a certain well defined current amplitude I, for instance, in magnetic devices for power applications immersed in the self-magnetic field. Consequently, we are able to define the proximity resistance per unit length by means of the relationship $P_{prox\,t,\,u.l.} = \frac{1}{2}R_{prox\,t,\,u.l.}I^2$:

$$R_{prox\,t,\,u.l.} = \frac{4\pi}{\delta}\Phi_{prox}(r_0/\delta)\bar{H}_t^2, \tag{19}$$

where $\bar{H}_t$ t is the normalized amplitude of the external transversal magnetic field $\left|H_t^0\right|$ divided by the total current I carried by the n_0 strands of the system.

The zeroth order TE mode in (2) is associated with a uniform longitudinal magnetic field H_z^0. TE modes are governed by the corresponding scalar equation with the structure of (9) substituting E_z by H_z. In this case, the first TE term solution has the following form [39]:

$$H_z(\mathbf{r}) = \frac{1}{\sigma}\frac{J_0\left((j-1)\rho/\delta\right)}{J_0\left((j-1)r_0/\delta\right)}H_z^0(\mathbf{r}_i) \quad \rho \leq r_0, \tag{20}$$

and

$$H_z(\mathbf{r}) = H_z^0(\mathbf{r}_i) \quad \rho > r_0. \tag{21}$$

The longitudinal magnetic field of the preceding equation can be associated with a transversal electric field inside the conductor by means of the identity $\mathbf{E}(\mathbf{r}) = \frac{1}{\sigma}\hat{\mathbf{z}} \times \nabla H_z(\mathbf{r})$ derived from the Maxwell's equations under the magneto-quasistatic approach $\nabla \times \mathbf{H}(\mathbf{r}) = \mathbf{J}(\mathbf{r})$, where the current density is $\sigma E(\mathbf{r})$. Thus, we have:

$$\mathbf{E}(\mathbf{r}) = \frac{j-1}{\sigma\delta}\frac{J_1\left((j-1)\rho/\delta\right)}{J_0\left((j-1)r_0/\delta\right)}H_z^0(\mathbf{r}_i)\hat{\varphi} \quad \rho \leq r_0, \tag{22}$$

The additional TE mode contribution $P_{prox\ z,\ u.l.}$ to the proximity power losses due to a uniform longitudinal external magnetic field $H_z^0(\mathbf{r}_i)$ can also be obtained integrating the factor $\frac{1}{2}\mathbf{E}\cdot\mathbf{J}^*$ in the volume of the unit length of the strand. Afterwards, the proximity resistance due to a longitudinal magnetic field $R_{prox\ z,\ u.l.}$ is obtained:

$$R_{prox\ z,\ u.l.} = \frac{2\pi}{\sigma}\Phi_{prox}(r_0/\delta)\bar{H}_z^2, \tag{23}$$

where $\bar{H}_z$ is the normalized amplitude of the external longitudinal magnetic field $|H_z^0(\mathbf{r}_i)|$ divided by the total current I carried by the system.

According to (19) and (23), proximity power losses due to transversal or longitudinal magnetic fields are characterized by the following aspects. The frequency dependence of the proximity resistance is accounted for by the ratio r_0/δ. The dependence functions are the same $\Phi_{prox(r0/\delta)}$ for either transversal or longitudinal magnetic fields. The proximity power losses for a transversal magnetic field doubles the proximity power losses for a longitudinal magnetic field of equal amplitude and frequency acting in a given strand, thus, $2R_{prox\ z,\ u.l.} = R_{prox\ t,\ u.l.}$ because the induced current densities in both cases are related by a $2\sin(\varphi - \varphi_0)$ factor. Moreover, owing to the orthogonality between the sources, the total proximity resistance can be calculated by the addition of the two contributions.

LOW- AND HIGH-FREQUENCY ANALYSIS

In many cases, it is worth obtaining simplified expressions that approach the exact function to evaluate the power losses based on extreme behavior of the system. The losses in the strand can be approached taking into account the dependence with respect to the frequency of the excitation, being possible to distinguish between the low frequency and the high frequency limit as is pointed out in [46] for different strand geometries. In the low frequency limit, the power losses are due to the external electromagnetic field disregarding the distortion introduced by the induced currents in the conductive medium. Note that the low frequency range is equivalent to small values of the ratio r_0/δ because the penetration depth of the strand δ has large values.

The conduction losses can be obtained considering a uniform longitudinal electric field $E_z^0\hat{z}$ equal to $(-V/l)\hat{z}$. In this case, the current density carried by the strand is $\sigma E_z^0\hat{z}$ uniformly distributed in the cross-section area, consequently the total current I_0 equals $\pi r_0^2 \sigma E_0$. Finally, the conduction resistance per unit length is obtained dividing the electromotive force into the total current, obtaining the well known dc conduction resistance per unit length $R_{cond\,u.l.}^{low}$ expressed as follows:

$$R_{cond\,u.l.}^{low} = \frac{1}{\pi r_0^2 \sigma}.$$

(24)

The previous result is equivalent to including the low frequency conduction factor $\Phi_{cond}^{low}(r_0/\delta)$ equal to one in (16).

The non-distorted electric field associated with an external transversal magnetic field $\mathbf{H}_t^0(\mathbf{r}_i)$ can be derived from (22) as:

$$\mathbf{E}_{\mathbf{H}_t^0}(\mathbf{r}) = j\omega\mu_0\rho\sin(\varphi - \varphi_0)H_t^0\hat{z}.$$

(25)

Therefore, the proximity losses arising from the preceding electric field can be calculated by integrating the power loss, thus:

$$P_{prox,\,u.l.}^{low} = \pi\omega^2\mu_0^2\sigma r_0^4 \bar{H}_t^2.$$

(26)

Rearranging the expression and applying the equivalence to the equivalent resistance, we obtain:

$$R_{prox\,t,\,u.l.}^{low} = \frac{4\pi}{\sigma}\Phi_{prox}^{low}(r_0/\delta)\bar{H}_t^2,$$

(27)

where we have:

$$\Phi_{prox}^{low}(r_0/\delta) = \frac{1}{4}\left(\frac{r_0}{\delta}\right)^4,$$

(28)

The values of $\Phi_{cond}^{low}(r_0/\delta)$ and $\Phi_{prox}^{low}(r_0/\delta)$ can also be obtained applying the small argument values to the complete functions shownin (16) and (18), respectively, assuming the identity $J_n(z) \cong \frac{1}{n!}\left(\frac{z}{2}\right)^n$ [47].

On the other hand, the high frequency limit power losses concern surface currents because electromagnetic fields exponentially decay inside the conductor for low values of δ with respect to the radius r_0, as is shown in [44, 45] for a flat half-space surface. Therefore, for a strand immersed in a uniform electric field $E_z^0(\mathbf{r}_i)\hat{\mathbf{z}}$, we have:

$$\mathbf{E}(\mathbf{r}) = E_z^0(\mathbf{r}_i)e^{-\frac{1+j}{\delta}\Delta}\hat{\mathbf{z}},\tag{29}$$

where Δ is the distance to the strand surface equal to $R_0 - \rho$. The current density J_c can be approached by a surface current density K_c provided by the following expression:

$$\mathbf{K}_c \cong \int_0^\infty \sigma E_z^0(\mathbf{r}_i)e^{-\frac{1+j}{\delta}\Delta}d\Delta\hat{\mathbf{z}},\tag{30}$$

hence:

$$\mathbf{K}_c \cong \frac{\sigma\delta}{1+j}E_z^0(\mathbf{r}_i)\hat{\mathbf{z}}.\tag{31}$$

As a result, the current I_0 carried by the strand is equal to:

$$I_0 \cong 2\pi r_0\frac{\sigma\delta}{1+j}E_z^0(\mathbf{r}_i).\tag{32}$$

Considering that $E_z^0(\mathbf{r}_i)$ is $-V/l$, and applying the identity $R = \Re e(V/I)$

$$R_{cond,\,u.l.}^{high} \cong \frac{1}{2\pi r_0\sigma\delta}.\tag{33}$$

In conclusion, we obtain the expression:

$$\Phi_{cond,\,u.l.}^{high}(r_0/\delta) = \frac{1}{2}\frac{r_0}{\delta}.\tag{34}$$

The factor $\Phi_{prox,\,u.l.}^{high}$. can be easily obtained starting with the configuration of the strand immersed in a uniform longitudinal magnetic field $H_z^0(\mathbf{r}_i)$ because the field is not distorted by the induced currents in the conductor. Additionally, considering the azimuthal electric field inside the strand with a dependence shown in (22) associated with a surface current similar to (31), together with the fact that the magnetic field vanishes inside the conductor, it is possible to define the so called impedance boundary condition [48–50] establishing the relationship between the tangential magnetic and electric field in the surface of the strand. In this case, it should be noted that both thelongitudinal electric field and the azimuthal electric field are tangential to the surface, thus:

$$Z_0 = \frac{E_\varphi(r_0)}{H_z^0(\mathbf{r}_i)} = \frac{1+j}{\sigma\delta}.$$

(35)

The electric field inside the strand also decays exponentially, as shown in (19). Consequently, the high frequency approach of the proximity power losses per surface unit $P_{prox\,z,\,u.s.}^{high}$ can be calculated by integrating the dissipated power density $\frac{1}{2}\mathbf{E}\cdot\mathbf{J}^*$ where the azimuthal electric field at the surface of the strand $E_\varphi(r_0)$ is $\frac{1+j}{\sigma\delta}H_z^0(\mathbf{r}_i)$. Thus, we have:

$$P_{prox\,z,\,u.s.}^{high} = \frac{1}{2}\int_0^\infty \frac{|H_z^0(\mathbf{r}_i)|^2}{\sigma\delta^2}e^{-\frac{2\Delta}{\delta}}d\Delta.$$

(36)

Integrating the above expression and multiplying by $2\pi r_0$ to calculate the power losses per unit length, we have:

$$P_{prox\,z,\,u.l.}^{high} = \frac{2\pi}{\sigma}\frac{r_0}{\delta}|H_z^0(\mathbf{r}_i)|^2.$$

(37)

As a result, the proximity factor at high frequencies is given by:

$$\Phi_{prox}^{high}(r_0/\delta) = \frac{r_0}{\delta}.$$

(38)

It should be noted that the expression for $\Phi_{cond}^{high}(r_0/\delta)$ and $\Phi_{prox}^{high}(r_0/\delta)$ can be derived from (16) and (18), respectively, considering the approximation for large argument values of the Bessel functions $J_n(z) \cong \sqrt{2/(\pi z)}\cos(z - \frac{1}{2}nz - \frac{1}{4}\pi)$ [47].

Figures 3(a) and 3(b) show the low- and high-frequency approaches compared with the complete expressions, respectively. The low frequency approach is accurate for the ratio up to one unit, whereas the high frequency approach can be used for ratios above several units.

EXPERIMENTAL VERIFICATION

The usefulness of the preceding expressions will be proven by being applied in a practical example consisting of the resistance calculation of a coil made with multi-stranded litz-wire. Two configurations are considered: first, an isolated coil placed in air and, second, a coil located above a magnetic flux concentrator which modifies the electromagnetic field, resulting in different resistance values. The total resistance is calculated by the addition of the different sources of power losses. Note that the ferrite layer acts as a current mirror increasing the magnetic field where the coil is immersed. As a consequence, the proximity losses increase.

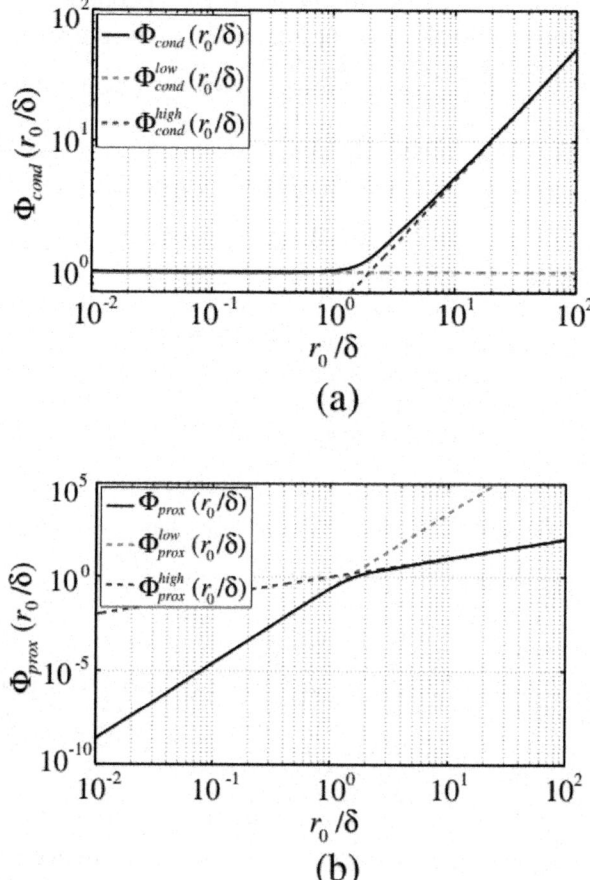

Figure 3: Comparison between complete expressions of conduction and proximity factors with respect to the low and high frequency approaches. (a) $\Phi_{cond}(r_0/\delta)$. (b) $\Phi_{prox}(r_0/\delta)$.

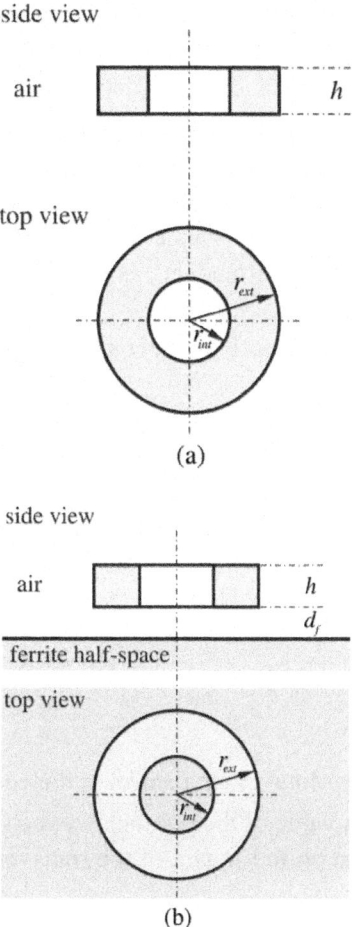

Figure 4: Geometrical structure of the two coil configurations experimentally measured. (a) Coil in air. (b) Coil above magnetic medium.

The measurements have been performed for a ring-type coil of n turns of wire composed of n_0 strands with a radius r_0 evenly distributed in the cross-section area of the coil. The geometrical characteristics of the circular coil are defined by the internal radius r_{int}, the external radius r_{ext}, and the height h, as shown in Figure 4(a). The coil is modeled as a uniform current density distribution due to the strand radius r_0 and the distance between the strands being smaller than the coil dimension. As a consequence, an ideal coil placed in air can be electrically characterized by its own frequency-independent inductance L. The total resistance of this type of device is taken into account by the addition of the equivalent resistances associated with the conduction

R_{cond} and proximity power losses R_{prox} in the winding.

Moreover, the second configuration is built adding a magnetic halfspace placed below the coil at a distance d_f, as depicted in Figure 4(b).

The conduction losses are associated with a conduction resistance value R_{cond} calculated considering that the strands in the bundle are connected in parallel. Consequently, the expression for R_{cond} is the wire length $n\pi(r_{ext} - r_{int})$ multiplied by the conduction resistance $R_{cond\ u.l.}$ per unit length of the strands and divided by the number of strands n_0, as is expressed as follows [19]:

$$R_{cond} = \frac{1}{r_0^2 \sigma}\frac{n}{n_0}\Phi_{cond}(r_0/\delta)(r_{ext} - r_{int}).$$
(39)

On the other hand, the proximity resistance R_{prox} is calculated by the addition of the losses arising from the external magnetic field acting over each strand. Note that the proximity losses are only originated by the transversal magnetic field because the longitudinal magnetic field is null due to the geometrical symmetry of the system. Consequently, the expression for the proximity resistivity is given by [19]:

$$R_{prox} = \frac{8\pi^2}{\sigma} n n_0 \Phi_{prox}(r_0/\delta)\left\langle r \cdot \bar{H}_0^2 \right\rangle,$$
(40)

where ρ is the radial coordinate with respect to the coil axis and $\left\langle \rho \cdot \bar{H}_0^2 \right\rangle$ is the coil cross-section mean value of the product between ρ and $\bar{H}_0^2$. Note that the proximity losses depend on the square of the transversal magnetic field along the strands.

The configuration measured in Figure 4(a) is a toroidal coil with internal radius r_{int} of 21.5 mm, external radius r_{ext} of 29 mm and height h of 4 mm. The coil is made winding 24 turns of litz-wire composed of 35 strands of radius r_0 equal to 75 μm of copper with electric conductivity σ at room temperature of 5.8 · 107 S/m. On the other hand, in Figure 4(b), the configuration consists of the preceding coil placed above a magnetic half-space at a distance d_f of 1 mm which is made of a high-magnetic ferrite with relative magnetic permeability μ_r of 2000.

The measurements have been performed by means of a precision LCR-meter Agilent E4940A at frequencies ranging from 1 kHz to

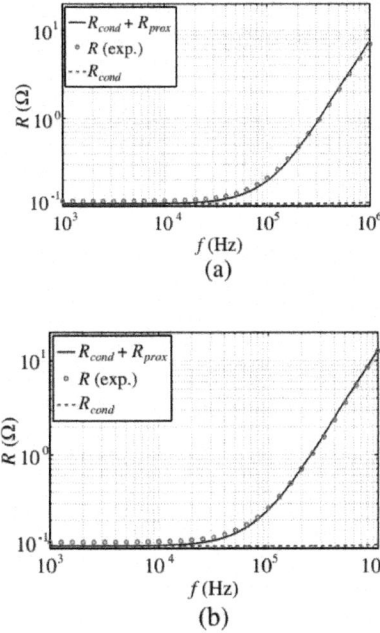

Figure 5: Experimental measurements and numerical results for a 24 turn coil in air and placed above a ferrite layer. Conduction resistances are also represented and the additional resistance is associated with proximity losses. (a) Coil in air. (b) Coil above ferrite layer.

1 MHz. Note that each frequency is associated with a penetration depth δ value of the strand. The output data is the equivalent impedance provided in equivalent resistance and inductance terms. Basically, the inductance depends on the system geometry having a weak relation with the internal structure of the wire. The inductance is almost constant in the overall frequency range in both cases, but is higher for the second configuration due to the presence of the magnetic layer. On the other hand, the resistance is due to the power losses in the winding because the dissipation in the remaining components of the system is negligible. Thus, the total resistance equals the sum of the conduction resistance R_{cond} and the proximity resistance R_{prox}.

Figures 5(a) and 5(b) show a good agreement between the experimental results and the numerical based values calculated with the addition of the conduction resistance obtained applying (39) and the proximity resistance from (40). In both cases, at the low frequency range, the conduction losses dominate. As a result, the value of the total resistance is essentially constant and no influence of the flux concentrator is observed because the conduction losses are independent of the magnetic field. On the other hand, at the high frequency range, the measured resistance is almost totally originated by the proximity losses. Thus, the total resistance greatly depends on the frequency and, additionally, the resistance is modified by the flux concentrator. Considering the previous analysis, the conduction resistance R_{cond} is almost constant in the whole frequency range, whereas the proximity resistance R_{prox} becomes the most important contribution above several tens of kHz.

CONCLUSION

The power losses in the winding of magnetic devices used in power systems made with multi-stranded litz-wire are analyzed in depth in this paper. A unified approach is developed for the different types of losses, either conduction or proximity losses. Moreover, external electromagnetic fields are considered as the sources modeling the effects of the complete system in the winding. The external electromagnetic field is decomposed into transversal magnetic TM and transversal electric TE components along the direction of the wire. The traditional power losses in the winding, called conduction and proximity losses, have been associated with a uniform longitudinal electric field and a transversal magnetic field, respectively. In this paper the relationship with the zeroth and first order TM modes have been established in order to achieve a unified analysis of the loss contributions. Additionally, the proximity losses due to the longitudinal magnetic field have been included by the action of the zeroth order TE mode. Both proximity loss contributions possess similar analytical expressions except for a proportional factor doubling the effect of the transversal magnetic field $\mathbf{H}_t^0(\mathbf{r}_i)$ with respect to the longitudinal magnetic field $H_z^0(\mathbf{r}_i)\hat{\mathbf{z}}$.

It is worth noting that the power losses in the winding are always related with the electric field of the analyzed mode because the losses are due to the electric currents in the strands. As a consequence, the external field originating the power losses in the winding can be substituted by an impressed current equivalent to the product between the conductivity of the strand and the electric field of the considered mode. As a result, the power losses can be easily evaluated. It should be noted that the evaluation at high frequencies includes the effect of the field distortion due to the conduction of the strand.

The analysis of power losses can be simplified taking into account the behavior of the system depending on the excitation frequency, which can be divided into a low- and high-frequency range depending on the value of the ratio r_0/δ. The low frequency ranges from dc to frequencies where r_0/δ is up to one unit, whereas at the high frequency limit r_0/δ is above several units. On the one hand, in the low frequency approach the external field is not distorted by the strand. On the other hand, in the high frequency approach the current densities in the strand become surface currents and the surface electric field and magnetic fields can be related by means of the so called impedance boundary conditions. It should be noted that simple expressions of power losses can be easily derived with an extended range of validity. Moreover, these approaches allow further analysis of more complex cross-section strand geometries where it is impractical to obtain exact analytical expressions.

Finally, the expressions obtained have been experimentally verified for two simple coil systems. The cases considered exhibit a high symmetric configuration, and therefore the power losses are only due to the TM modes. However, the power losses in less symmetric systems, for instance transformers or inductors used in commercial domestic heaters, should be accounted for by considering the additional TE contribution originated by the longitudinal magnetic field $H_z^0(\mathbf{r}_i)\hat{\mathbf{z}}$.

ACKNOWLEDGMENT

This work was partly supported by the Spanish MICINN under Project CSD2009-00046, Project TEC2010-19207, and Project IPT-2011-1158-920000, and by the Bosch and Siemens Home Appliances Group.

REFERENCES

1. Lammeraner, J. and M. Stafl, Eddy Currents, Chemical Rubber Co., Cleveland, Ohio, 1964.

2. Urling, A. M., et al., "Characterizing high-frequency effects in transformer windings — A guide to several significant articles," Applied Power Electronics Conference, 373–385, Baltimore, USA, 1989.

3. Reatti, A. and M. K. Kazimierczuk, "Comparison of various methods for calculating the AC resistance of inductors," IEEE Transactions on Magnetics, Vol. 38, No. 3, 1512–1518, 2002.

4. Nan, X. and C. R. Sullivan, "An improved calculation of proximity-effect loss in high-frequency windings of round conductors," Power Electronics Specialist Conference, 853–860, Acapulco, Mexico, 2003.

5. Dowell, P. L., "Effects of eddy currents in transformer windings," Proceedings of the Institution of Electrical Engineers, Vol. 113, No. 8, 1387–1394, 1966.

6. Robert, F., "A theoretical discussion about the layer copper factor used in winding losses calculation," IEEE Transactions on Magnetics, Vol. 38, No. 5, 3177–3179, 2002.

7. Sippola, M. and R. E. Sepponen, "Accurate prediction of highfrequency

power-transformer losses and temperature rise," IEEE Transactions on Power Electronics, Vol. 17, No. 5, 835–847, 2002.

8. Acero, J., et al., "A model of losses in twisted-multistranded wires for planar windings used in domestic induction heating appliances," Applied Power Electronics Conference, 1247–1253, Anaheim, USA, 2007.

9. Lotfi, A. W. and F. C. Lee, "A high frequency model for Litz wire for switch-mode magnetics," Industry Applications Society Annual Meeting, 1169–1175, Toronto, Canada, 1993.

10. Ferreira, J. A., "Analytical computation of AC resistance of round and rectangular litz wire windings," IEE Proceedings B, Electric Power Applications, Vol. 139, No. 1, 21–25, 1992.

11. Lotfi, A. W., P. M. Gradzki, and F. C. Lee, "Proximity effects in coils for high frequency power applications," IEEE Transactions on Magnetics, Vol. 28, No. 5, 2169–2171, 1992.

12. Albach, M., "Two-dimensional calculation of winding losses in transformers," Power Electronics Specialists Conference, 1639– 1644, Galway, Ireland, 2000.

13. Tourkhani, F. and P. Viarouge, "Accurate analytical model of winding losses in round Litz wire windings," IEEE Transactions on Magnetics, Vol. 37, No. 1, 538–543, 2001.

14. Spang, M. and M. Albach, "Optimized winding layout for minimized proximity losses in coils with rod cores," IEEE Transactions on Magnetics, Vol. 44, No. 7, 1815–1821, 2008.

15. Larouci, C., et al., "Copper losses of flyback transformer: Search for analytical expressions," IEEE Transactions on Magnetics, Vol. 39, No. 3, 1745–1748, 2003.

16. Kazimierczuk, M. K., High-frequency Magnetic Components, John Wiley & Sons Ltd, Chichester, UK, 2009.

17. Perry, M. P., "On calculating losses in current carrying conductors in an external alternating magnetic field," IEEE Transactions on Magnetics, Vol. 17, No. 5, 2486–2488, 1981.

18. Fawzi, T. H., P. E. Burke, and B. R. McLean, "Eddy losses and power shielding of cylindrical shells in transverse and axial magnetic fields," IEEE Transactions on Magnetics, Vol. 31, No. 3, 1452–1455, 1995.

19. Carretero, C., R. Alonso, J. Acero, O. Lucia, and J. M. Burdio, "Dissipative losses evaluation in magnetic power devices with litzwire type windings," PIERS Online, Vol. 7, No. 3, 246–250, 2011.

20. Namjoshi, K. V., J. D. Lavers, and P. P. Biringer, "Eddy current power

loss in structural steel due to cables carrying current in a perpendicular direction," IEEE Transactions on Magnetics, Vol. 30, No. 1, 85–91, 1994.

21. Sullivan, C. R., "Optimal choice for number of strands in a litz-wire transformer winding," IEEE Transactions on Power Electronics, Vol. 14, No. 2, 283–291, 1999.

22. Sullivan, C. R., "Computationally efficient winding loss calculation with multiple windings, arbitrary waveforms, and twodimensional or three-dimensional field geometry," IEEE Transactions on Power Electronics, Vol. 16, No. 1, 142–150, 2001.

23. Nan, X. and C. R. Sullivan, "Simplified high-accuracy calculation of eddy-current loss in round-wire windings," Power Electronics Specialists Conference, 873–879, Aachen, Germany, 2004.

24. Koertzen, H. W. E., J. D. van Wyk, and J. A. Ferreira, "An investigation of the analytical computation of inductance and AC resistance of the heat-coil for induction cookers," Industry Applications Society Conference, 1113–1119, Houston, USA, 1992.

25. Acero, J., et al., "Frequency-dependent resistance in litz wire planar windings for domestic induction heating appliances," IEEE Transactions on Power Electronics, Vol. 21, No. 4, 856–866, 2006.

26. Hernandez, P., et al., "Power losses distribution in the litzwire winding of an inductor for an induction cooking appliance," Conference of the Industrial Electronics Society, 1134–1137, Sevilla, Spain, 2002.

27. Podoltsev, A. D., I. N. Kucheryavaya, and B. B. Lebedev, "Analysis of effective resistance and eddy-current losses in multiturn winding of high-frequency magnetic components," IEEE Transactions on Magnetics, Vol. 39, No. 1, 539–548, 2003.

28. Dular, P. and J. Gyselinck, "Modeling of 3-D stranded inductors with the magnetic vector potential formulation and spatially dependent turn voltages of reduced support," IEEE Transactions on Magnetics, Vol. 40, No. 2, 1298–1301, 2004.

29. Gyselinck, J. and P. Dular, "Frequency-domain homogenization of bundles of wires in 2-D magnetodynamic FE calculations," IEEE Transactions on Magnetics, Vol. 41, No. 5, 1416–1419, 2005.

30. Gyselinck, J., R. V. Sabariego, and P. Dular, "Time-domain homogenization of windings in 2-D finite element models," IEEE Transactions on Magnetics, Vol. 43, No. 4, 1297–1300, 2007.

31. Sabariego, R. V., P. Dular, and J. Gyselinck, "Time-domain homogenization of windings in 3-D finite element models," IEEE

Transactions on Magnetics, Vol. 44, No. 6, 1302–1305, 2008.

32. Kong, J. A., "Electromagnetic fields due to dipole antennas over stratified anisotropic media," Geophysics, Vol. 37, No. 6, 985–996, 1972.

33. Clemmow, P. C., "The resolution of a dipole field into transverse electric and transverse magnetic waves," Proceedings of the Institution of Electrical Engineers, Vol. 110, No. 1, 107–111, 1963.

34. Wilton, D., "A TM-TE decomposition of the electromagnetic field due to arbitrary sources radiating in unbounded regions," IEEE Transactions on Antennas and Propagation, Vol. 28, No. 1, 111– 114, 1980.

35. Lindell, I. V., "TE/TM decomposition of electromagnetic sources," IEEE Transactions on Antennas and Propagation, Vol. 36, No. 10, 1382–1388, 1988.

36. Weiss, S. J. and W. K. Kahn, "Decomposition of electromagnetic boundary conditions at planar interfaces with applications to TE and TM field solutions," IEEE Transactions on Antennas and Propagation, Vol. 46, No. 11, 1687–1691, 1998.

37. Janaswamy, R., "A note on the TE/TM decomposition of electromagnetic fields in three dimensional homogeneous space," IEEE Transactions on Antennas and Propagation, Vol. 52, No. 9, 2474–2477, 2004.

38. Fawzi, T. H. and P. E. Burke, "Use of surface integral equations for analysis of TM-induction problem," Proceedings of the Institution of Electrical Engineers, Vol. 121, No. 10, 1109–1116, 1974.

39. Fawzi, T. H., P. E. Burke, and M. Fabiano-Alves, "Use of surfaceintegral equations for the analysis of the TE-induction problem," Proceedings of the Institution of Electrical Engineers, Vol. 123, No. 7, 725–728, 1976.

40. Carretero, C., R. Alonso, J. Acero, and J. M. Burdio, "Coupling impedance between planar coils inside a layered media," Progress In Electromagnetics Research, Vol. 112, 381–396, 2011.

41. Carcione, J. M., "Simulation of electromagnetic diffusion in anisotropic media," Progress In Electromagnetics Research B, Vol. 26, 425–450, 2010.

42. Rothwell, E. J. and M. J. Cloud, Electromagnetics, CRC Press, Boca Raton, 2000.

43. Ferreira, J. A., "Improved analytical modeling of conductive losses in magnetic components," IEEE Transactions on Power Electronics, Vol. 9, No. 1, 127–131, 1994.

44. Silveira, F. E. M. and J. A. S. Lima, "Skin effect from extended irreversible thermodynamics perspective," Journal of Electromagnetic Waves and

Applications, Vol. 24, Nos. 2–3, 151– 160, 2010.

45. Voyer, D., R. Perrusel, and P. Dular, "Perturbation method for the calculation of losses inside conductors in microwave structures," Progress In Electromagnetics Research, Vol. 103, 339– 354, 2010.

46. Burke, P., T. Fawzi, and T. Akinbiyi, "The use of asymptotes to estimate TE- and TM-mode losses in long conductors," IEEE Transactions on Magnetics, Vol. 14, No. 5, 374–376, 1978.

47. Abramowitz, M. and I. A. Stegun, Handbook of Mathematical Functions: With Formulas, Graphs, and Mathematical Tables, U.S. Dept. of Commerce, Washington, D.C., 1970.

48. Qian, Z.-G., M.-S. Tong, and W. C. Chew, "Conductive medium modeling with an augmented GIBC formulation," Progress In Electromagnetics Research, Vol. 99, 261–272, 2009.

49. Fawzi, T., M. Ahmed, and P. Burke, "On the use of the impedance boundary conditions in eddy current problems," IEEE Transactions on Magnetics, Vol. 21, No. 5, 1835–1840, 1985.

50. Yuferev, S. and N. Ida, Surface Impedance Boundary Conditions. A Comprehensive Approach, CRC Press, Boca Raton, 2009.

Chapter 5

POWER ELECTRONICS

INTRODUCTION

Power electronics makes up a large part of engineering and has close connections with many areas of physics, chemistry, and mechanics. It establishes a rapidly expanding field in electrical engineering and a scope of its technology covers a wide spectrum. Power applications with electronic converters do a lot of difficult work for us. Optimists envision power electronics doing more and more things for the population. Electronic appliances contribute to a healthier and more comfortable live the world over. Thanks to advances in science and related technology, many people no longer have to spend much time working for the bare necessities of life. Whatever it is that we really want to do, power electronics helps us to do it better.

In the first half of the 20th century, electronic equipment was mainly based on vacuum tubes, such as gas-discharge valves, thyratrons, mercury arc rectifiers, and ignitrons. Until the end of the 1920th, vacuum diodes (kenotrones) were the main electronic devices. In the 1930th, they were replaced by mercury equipment. The majority of valves were arranged as coaxial closed cylinders round the cathode. Valves that were more complex contained several gridded electrodes between the cathode and anode. The vacuum tube had a set of disadvantages. First, it had an internal power heater. Second, its life was limited by a few thousand hours before its filament burns out. Third, it was bulky. Fourth, it gave off heat that raised the internal temperature of the electronics equipment. Because of vacuum tube technology, the first electronic devices were very expensive and dissipated a great deal of power.

Today, power electronics is a rapidly expanding field in electrical engineering and a scope of the technology covers a wide spectrum of electronic converters. Different kinds of power supplies are used everywhere in normal daily routines both at home, office and industry. This is due to the progress in electronic components and equipment development that has been achieved in the last few decades. Electronic and electrical apparatus are everywhere, and all these devices need electrical power to work. Most of electronic supplies are switching semiconductor converters thanks to the efficiency, size, capability to operate at various current and voltage levels, control features and price compared to the linear power supply.

POWER SEMICONDUCTOR DEVICES

The modern age of power electronics began with the introduction of thyristors in the late 1950s. Now there are several types of power devices available for high-power and high-frequency applications. The most notable power devices are gate turn-off thyristors, power Darlington transistors, power MOSFETs, and insulated-gate bipolar transistors (IGBTs). Power semiconductor devices are the most important functional elements in all power conversion applications. The power devices are mainly used as switches to convert power from one form to another. They are used in motor control systems, uninterrupted power supplies, high-voltage dc transmission, power supplies, induction heating, and in many other power conversion applications. A review of the basic characteristics of these power devices is presented in this section.

Thyristor and Triac

The thyristor, also called a silicon-controlled rectifier (SCR), is basically a four-layer three-junction pnpn device. It has three terminals: anode, cathode, and gate. The device is turned on by applying a short pulse across the gate and cathode. Once the device turns on, the gate loses its control to turn off the device. The turn-off is achieved by applying a reverse voltage across the anode and cathode. The thyristor symbol and its volt-ampere characteristics are shown in Fig. 1. There are basically two classifications of thyristors: converter grade and inverter grade. The difference between a converter-grade and an inverter-grade thyristor is the low turn-off time (on the order of a few microseconds) for the latter. The converter-grade thyristors are slow type and are used in natural commutation (or phase-controlled) applications. Inverter-grade thyristors are used in forced commutation applications such as dc-dc choppers and dc-ac inverters. The inverter-grade thyristors are turned off by forcing the current to zero using an external commutation circuit. This requires additional commutating components, thus resulting in additional losses in the inverter.

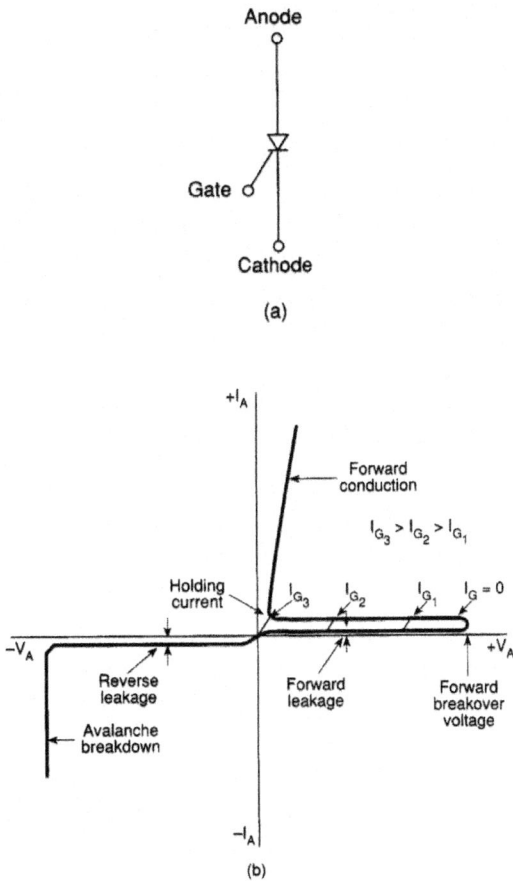

Figure 1: (a) Thyristor symbol and (b) volt-ampere characteristics. (Source: B.K. Bose, Modern Power Electronics: Evaluation, Technology, and Applications, p. 5. © 1992 IEEE.).

Thyristors are highly rugged devices in terms of transient currents, di/dt, and dv/dt capability. The forward voltage drop in thyristors is about 1.5 to 2 V, and even at higher currents of the order of 1000 A, it seldom exceeds 3 V. While the forward voltage determines the on-state power loss of the device at any given current, the switching power loss becomes a dominating factor affecting the device junction temperature at high operating frequencies. Because of this, the maximum switching frequencies possible using thyristors are limited in comparison with other power devices considered in this section.

Thyristors have I^2t withstand capability and can be protected by fuses. The nonrepetitive surge current capability for thyristors is about 10 times their rated root mean square (rms) current. They must be protected by snubber networks for dv/dt and di/dt effects. If the specified dv/dt is exceeded, thyristors may start conducting without applying a gate pulse. In dc-to-ac conversion applications it is necessary to use an antiparallel diode of similar rating across each main thyristor. Thyristors are available up to 6000 V, 3500 A.

A triac is functionally a pair of converter-grade thyristors connected in antiparallel. The triac symbol and volt-ampere characteristics are shown in Fig. 2. Because of the integration, the triac has poor reapplied dv/dt, poor gate current sensitivity at turn-on, and longer turn-off time. Triacs are mainly used in phase control applications such as in ac regulators for lighting and fan control and in solid-state ac relays.

Gate Turn-Off Thyristor (GTO)

The GTO is a power switching device that can be turned on by a short pulse of gate current and turned off by a reverse gate pulse. This reverse gate current amplitude is dependent on the anode current to be turned off. Hence there is no need for an external commutation circuit to turn it off. Because turn-off is provided by bypassing carriers directly to the gate circuit, its turn-off time is short, thus giving it more capability for highfrequency operation than thyristors. The GTO symbol and turn-off characteristics are shown in Fig. 3.

GTOs have the I²t withstand capability and hence can be protected by semiconductor fuses. For reliable operation of GTOs, the critical aspects are proper design of the gate turn-off circuit and the snubber circuit.

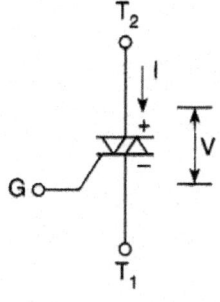

(a)

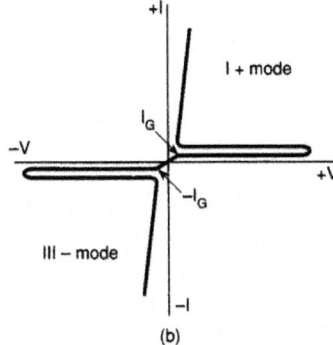

(b)

Figure 2: (a) Triac symbol and (b) volt-ampere characteristics. (Source: B.K. Bose, Modern Power Electronics: Evaluation, Technology, and Applications, p. 5. © 1992 IEEE.).

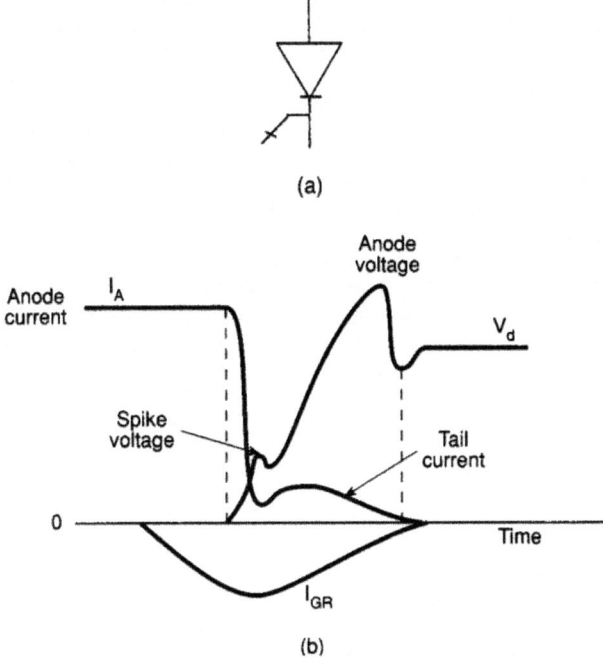

(a)

(b)

Figure 3: (a) GTO symbol and (b) turn-off characteristics. (Source: B.K. Bose, Modern Power Electronics: Evaluation, Technology, and Applications, p. 5. © 1992 IEEE.).

A GTO has a poor turn-off current gain of the order of 4 to 5. For example, a 2000-A peak current GTO may require as high as 500 A of reverse gate current. Also, a GTO has the tendency to latch at temperatures above 125°C. GTOs are available up to about 4500 V, 2500 A.

Reverse-Conducting Thyristor (RCT) and Asymmetrical Silicon-Controlled Rectifier (ASCR)

Normally in inverter applications, a diode in antiparallel is connected to the thyristor for commutation/freewheeling purposes. In RCTs, the diode is integrated with a fast switching thyristor in a single silicon chip. Thus, the number of power devices could be reduced. This integration brings forth a substantial improvement of the static and dynamic characteristics as well as its overall circuit performance.

The RCTs are designed mainly for specific applications such as traction drives. The antiparallel diode limits the reverse voltage across the thyristor to 1 to 2 V. Also, because of the reverse recovery behavior of the diodes, the thyristor may see very high reapplied dv/dt when the diode recovers from its reverse voltage. This necessitates use of large RC snubber networks to suppress voltage transients. As the range of application of thyristors and diodes extends into higher frequencies, their reverse recovery charge becomes increasingly important. High reverse recovery charge results in high power dissipation during switching.

The ASCR has a similar forward blocking capability as an inverter-grade thyristor, but it has a limited reverse blocking (about 20–30 V) capability. It has an on-state voltage drop of about 25% less than an inverter-grade thyristor of a similar rating. The ASCR features a fast turn-off time; thus it can work at a higher frequency than an SCR. Since the turn-off time is down by a factor of nearly 2, the size of the commutating components can be halved. Because of this, the switching losses will also be low.

Gate-assisted turn-off techniques are used to even further reduce the turn-off time of an ASCR. The application of a negative voltage to the gate during turn-off helps to evacuate stored charge in the device and aids the recovery mechanisms. This will in effect reduce the turn-off time by a factor of up to 2 over the conventional device.

Power Transistor

Power transistors are used in applications ranging from a few to several hundred kilowatts and switching frequencies up to about 10 kHz. Power transistors used in power conversion applications are generally npn type. The power transistor is turned on by supplying sufficient base current, and this base drive has to be maintained throughout its conduction period. It is turned off by removing the base drive and making the base voltage slightly negative (within $-V_{BE(max)}$). The saturation voltage of the device is normally 0.5 to 2.5 V and increases as the current increases. Hence the on-state losses increase more than proportionately with current. The transistor off-state losses are much lower than the on-state losses because the leakage current of the device is of the order of a few milliamperes. Because of relatively larger switching times, the switching loss significantly increases with switching frequency. Power transistors can block only forward voltages. The reverse peak voltage rating of these devices is as low as 5 to 10 V.

Power transistors do not have I^2t withstand capability. In other words, they can absorb only very little energy before breakdown. Therefore, they cannot be protected by semiconductor fuses, and thus an electronic protection method has to be used.

To eliminate high base current requirements, Darlington con- figurations are commonly used. They are available in monolithic or in isolated packages. The basic Darlington configuration is shown schematically in Fig. 4. The Darlington configuration presents a specific advantage in that it can considerably increase the current switched by the transistor for a given base drive. The $V_{CE(sat)}$ for the Darlington is generally more than that of a single transistor of similar rating with corresponding increase in onstate power loss. During switching, the reverse-biased collector junction may show hot spot breakdown effects that are specified by reverse-bias safe operating area (RBSOA) and forward bias safe operating area (FBSOA). Modern devices with highly interdigited emitter base geometry force more uniform current distribution and therefore considerably improve second breakdown effects. Normally, a well-designed switching aid network constrains the device operation well within the SOAs.

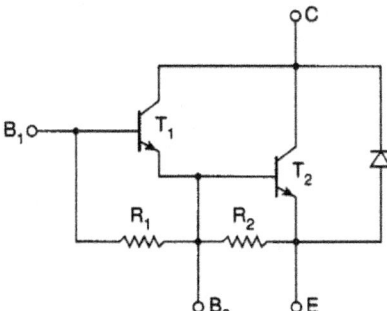

Figure 4: A two-stage Darlington transistor with bypass diode. (Source: B.K. Bose, Modern Power Electronics: Evaluation, Technology, and Applications, p. 6. © 1992 IEEE.).

Power MOSFET

Power MOSFETs are marketed by different manufacturers with differences in internal geometry and with different names such as MegaMOS, HEXFET, SIPMOS, and TMOS. They have unique features that make them potentially attractive for switching applications. They are essentially voltage-driven rather than current-driven devices, unlike bipolar transistors.

The gate of a MOSFET is isolated electrically from the source by a layer of silicon oxide. The gate draws only a minute leakage current of the order of nanoamperes. Hence the gate drive circuit is simple and power loss in the gate control circuit is practically negligible. Although in steady state the gate draws virtually no current, this is not so under transient conditions. The gate-to-source and gate-to-drain capacitances have to be charged and discharged appropriately to obtain the desired switching speed, and the drive circuit must have a sufficiently low output impedance to supply the required charging and discharging currents. The circuit symbol of a power MOSFET is shown in Fig. 5.

Power MOSFETs are majority carrier devices, and there is no minority carrier storage time. Hence they have exceptionally fast rise and fall times. They are essentially resistive devices when turned on, while bipolar transistors present a more or less constant $V_{CE(sat)}$ over the normal operating range. Power dissipation in MOSFETs is $I_d^2 R_{DS(on)}$, and in bipolars it is $I_C V_{CE(sat)}$. At low currents, therefore, a power MOSFET may have a lower conduction loss than a comparable bipolar device, but at higher currents, the conduction loss will exceed that of bipolars. Also, the $R_{DS(on)}$ increases with temperature.

An important feature of a power MOSFET is the absence of a secondary breakdown effect, which is present in a bipolar transistor, and as a result, it has an extremely rugged switching performance. In MOSFETs, $R_{DS(on)}$ increases with temperature, and thus the current is automatically diverted away from the hot spot. The drain body junction appears as an antiparallel diode between source and drain. Thus power MOSFETs will not support voltage in the reverse direction. Although this inverse diode is relatively fast, it is slow by comparison with the MOSFET. Recent devices have the diode recovery time as low as 100 ns. Since MOSFETs cannot be protected by fuses, an electronic protection technique has to be used.

With the advancement in MOS technology, ruggedized MOSFETs are replacing the conventional MOSFETs. The need to ruggedize power MOSFETs is related to device reliability. If a MOSFET is operating within its specification range at all times, its chances for failing catastrophically are minimal. However, if its absolute maximum rating is exceeded, failure probability increases dramatically. Under actual operating conditions, a MOSFET may be subjected to transients — either externally from the power bus supplying the circuit or from the circuit itself due, for example, to inductive kicks going beyond the absolute maximum ratings. Such conditions are likely in almost every application, and in most cases are beyond a designer's control. Rugged devices are made to be more tolerant for over-voltage transients. Ruggedness is the ability of a MOSFET to operate in an environment of dynamic electrical stresses, without activating any of the parasitic bipolar junction transistors. The rugged device can withstand higher levels of diode recovery dv/dt and static dv/dt.

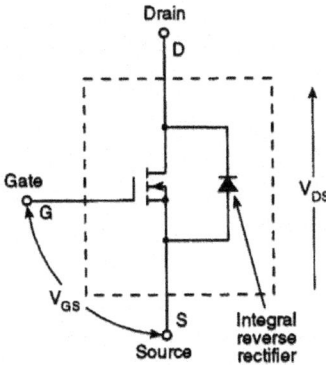

Figure 5: Power MOSFET circuit symbol. (Source: B.K. Bose, Modern Power Electronics: Evaluation, Technology, and Applications, p. 7. © 1992 IEEE.).

Insulated-Gate Bipolar Transistor (IGBT)

The IGBT has the high input impedance and high-speed characteristics of a MOSFET with the conductivity characteristic (low saturation voltage) of a bipolar transistor. The IGBT is turned on by applying a positive voltage between the gate and emitter and, as in the MOSFET, it is turned off by making the gate signal zero or slightly negative. The IGBT has a much lower voltage drop than a MOSFET of similar ratings. The structure of an IGBT is more like a thyristor and MOSFET. For a given IGBT, there is a critical value of collector current that will cause a large enough voltage drop to activate the thyristor. Hence, the device manufacturer specifies the peak allowable collector current that can flow without latch-up occurring. There is also a corresponding gate source voltage that permits this current to flow that should not be exceeded.

Like the power MOSFET, the IGBT does not exhibit the secondary breakdown phenomenon common to bipolar transistors. However, care should be taken not to exceed the maximum power dissipation and specified maximum junction temperature of the device under all conditions for guaranteed reliable operation. The onstate voltage of the IGBT is heavily dependent on the gate voltage. To obtain a low on-state voltage, a sufficiently high gate voltage must be applied.

In general, IGBTs can be classified as punchthrough (PT) and nonpunch-through (NPT) structures, as shown in Fig. 6. In the PT IGBT, an N^+ buffer layer is normally introduced between the P^+ substrate and the N^- epitaxial layer, so that the whole N^- drift region is depleted when the device is blocking the off-state voltage, and the electrical field shape inside the N^- drift region is close to a rectangular shape. Because a shorter N^- region can be used in the punch-through IGBT, a better trade-off between the forward voltage drop and turn-off time can be achieved. PT IGBTs are available up to about 1200 V.

High voltage IGBTs are realized through nonpunch-through process. The devices are built on a N^- wafer substrate which serves as the N^- base drift region. Experimental NPT IGBTs of up to about 4 KV have been reported in the literature. NPT IGBTs are more robust than PT IGBTs particularly under short circuit conditions. But NPT IGBTs have a higher forward voltage drop than the PT IGBTs.

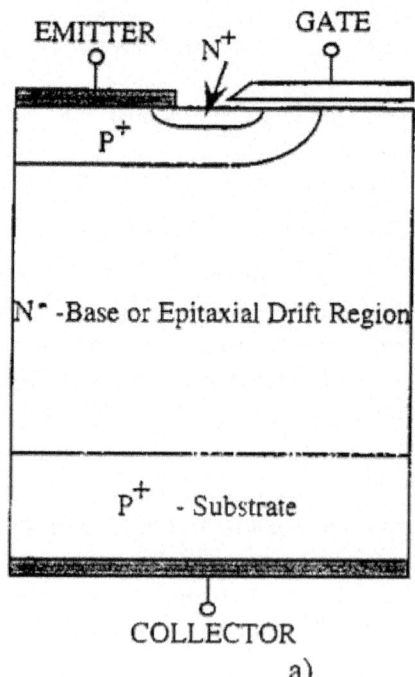

a)

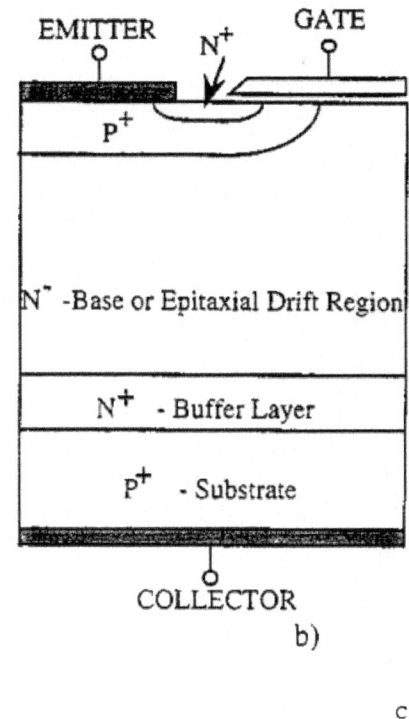

EMITTER   N⁺   GATE

P⁺

N⁻ -Base or Epitaxial Drift Region

N⁺ - Buffer Layer

P⁺ - Substrate

COLLECTOR

b)

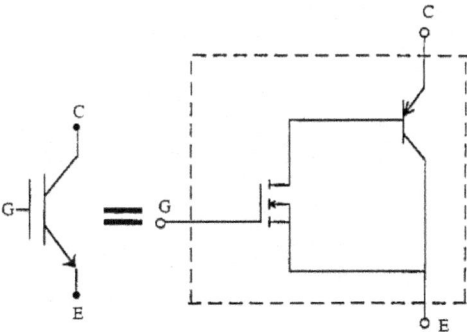

Figure 6: Nonpunch-through IGBT, (b) Punchthrough IGBT, (c) IGBT equivalent circuit.

The PT IGBTs cannot be as easily paralleled as MOSFETs. The factors that inhibit current sharing of parallel-connected IGBTs are (1) on-state current unbalance, caused by $V_{CE(sat)}$ distribution and main circuit wiring resistance distribution, and (2) current unbalance at turn-on and turn-off, caused by the switching time difference of the parallel connected devices and circuit wiring inductance distribution. The NPT IGBTs can be paralleled because of their positive temperature coefficient property.

MOS-Controlled Thyristor (MCT)

The MCT is a new type of power semiconductor device that combines the capabilities of thyristor voltage and current with MOS gated turn-on and turn-off. It is a high power, high frequency, low conduction drop and a rugged device, which is more likely to be used in the future for medium and high power applications. A cross sectional structure of a p-type MCT with its circuit schematic is shown in Fig. 7. The MCT has a thyristor type structure with three junctions and PNPN layers between the anode and cathode. In a practical MCT, about 100,000 cells similar to the one shown are paralleled to achieve the desired current rating. MCT is turned on by a negative voltage pulse at the gate with respect to the anode, and is turned off by a positive voltage pulse.

The MCT was announced by the General Electric R & D Center on November 30, 1988. Harris Semiconductor Corporation has developed two generations of p-MCTs. Gen-1 p-MCTs are available at 65 A/1000 V and 75A/600 V with peak controllable current of 120 A. Gen-2 p-MCTs are being developed at similar current and voltage ratings, with much improved turn-on capability and switching speed. The reason for developing p-MCT is the fact that the current density that can be turned off is 2 or 3 times higher than that of an n-MCT; but n-MCTs are the ones needed for many practical applications. Harris Semiconductor Corporation is in the process of developing n-MCTs, which are expected to be commercially available during the next one to two years.

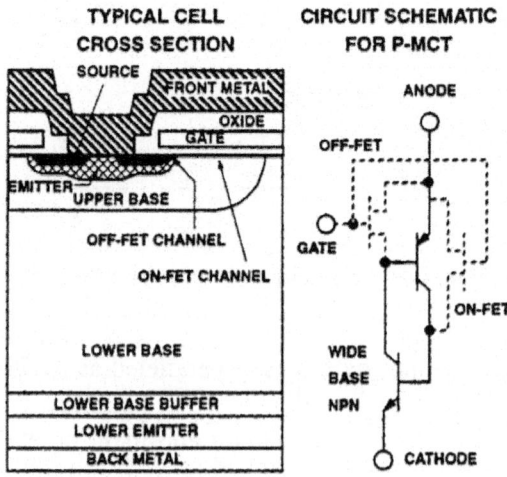

Figure 7: (Source: Harris Semiconductor, User's Guide of MOS Controlled Thyristor, With permission.).

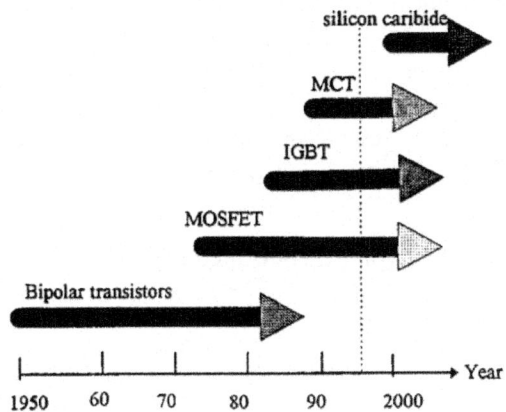

Figure 8: Current and future pwer semiconductor devices development direction (Source: A.Q. Huang, Recent Developments of Power Semiconductor Devices, VPEC Seminar Proceedings, pp. 1–9. With permission.).

The advantage of an MCT over-IGBT is its low forward voltage drop. N-type MCTs will be expected to have a similar forward voltage drop, but with an improved reverse bias safe operating area and switching speed. MCTs have relatively low switching times and storage time. The MCT is capable of high current densities and blocking voltages in both directions. Since the power gain of an MCT is extremely high, it could be driven directly from logic gates.An MCT has high di/dt (of the order of 2500 A/ms) and high dv/dt (of the order of 20,000 V/ms) capability. The MCT, because of its superior characteristics, shows a tremendous possibility for applications such as motor drives, uninterrupted power supplies, static VAR compensators, and high power active power line conditioners.

The current and future power semiconductor devices developmental direction is shown in Fig. 8. High temperature operation capability and low forward voltage drop operation can be obtained if silicon is replaced by silicon carbide material for producing power devices. The silicon carbide has a higher band gap than silicon. Hence higher breakdown voltage devices could be developed. Silicon carbide devices have excellent switching characteristics and stable blocking voltages at higher temperatures. But the silicon carbide devices are still in the very early stages of development.

POWER CONVERSION

Power conversion deals with the process of converting electric power from one form to another. The power electronic apparatuses performing the power conversion are called power converters. Because they contain no moving parts, they are often referred to as static power converters. The power conversion is achieved using power semiconductor devices, which are used as switches. The power devices used are SCRs (silicon controlled rectifiers, or thyristors), triacs, power transistors, power MOSFETs, insulated gate bipolar transistors (IGBTs), and MCTs (MOS-controlled thyristors). The power converters are generally classified as:

1. ac-dc converters (phase-controlled converters)

2. direct ac-ac converters (cycloconverters)

3. dc-ac converters (inverters)

4. dc-dc converters (choppers, buck and boost converters)

AC-DC Converters

The basic function of a phase-controlled converter is to convert an alternating voltage of variable amplitude and frequency to a variable dc voltage. The power devices used for this application are generally SCRs. The average value of the output voltage is controlled by varying the conduction time of the SCRs. The turn-on of the SCR is achieved by providing a gate pulse when it is forward-biased. The turn-off is achieved by the commutation of current from one device to another at the instant the incoming ac voltage has a higher instantaneous potential than that of the outgoing wave. Thus there is a natural tendency for current to be commutated from the outgoing to the incoming SCR, without the aid of any external commutation circuitry. This commutation process is often referred to as natural commutation. A single-phase half-wave converter is shown in Fig. 9. When the SCR is turned on at an angle a, full supply voltage (neglecting the SCR drop) is applied to the load. For a purely resistive load, during the positive half cycle, the output voltage waveform follows the input ac voltage waveform. During the negative half cycle, the SCR is turned off. In the case of inductive load, the energy stored in the inductance causes the current to flow in the load circuit even after the reversal of the supply voltage, as shown in Fig. 9(b). If there is no freewheeling diode DF , the load current is discontinuous. A freewheeling diode is connected across the load to turn off the SCR as soon as the input voltage polarity reverses, as shown in Fig. 9(c). When the SCR is off, the load current will freewheel through the diode. The power flows from the input to the load only when the SCR is conducting. If there is no freewheeling diode, during the negative portion of the supply voltage, SCR returns the energy stored in the load inductance to the supply. The freewheeling diode improves the input power factor.

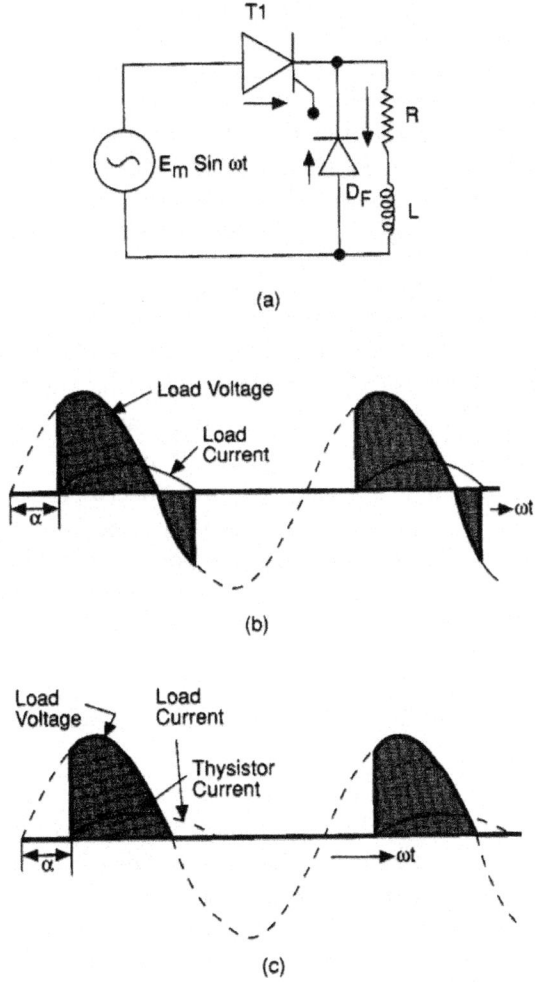

(a)

(b)

(c)

Figure 9: Single-phase half-wave converter with freewheeling diode. (a) Circuit diagram; (b) waveform for inductive load with no freewheeling diode; (c) waveform with freewheeling diode.

The controlled full-wave dc output may be obtained by using either a center tap transformer (Fig. 10) or by bridge configuration (Fig. 11). The bridge configuration is often used when a transformer is undesirable and the magnitude of the supply voltage properly meets the load voltage requirements. The average output voltage of a single-phase full-wave converter for continuous current conduction is given by

$$v_{d\alpha} = 2\frac{E_m}{\pi}\cos\alpha$$

where E_m is the peak value of the input voltage and a is the firing angle. The output voltage of a single-phase bridge circuit is the same as that shown in Fig. 10. Various configurations of the single-phase bridge circuit can be obtained if, instead of four SCRs, two diodes and two SCRs are used, with or without freewheeling diodes.

A three-phase full-wave converter consisting of six thyristor switches is shown in Fig. 12(a). This is the most commonly used three-phase bridge configuration. Thyristors T_1, T_3, and T_5 are turned on during the positive half cycle of the voltages of the phases to which they are connected, and thyristors T_2, T_4, and T_6 are turned on during the negative half cycle of the phase voltages. The reference for the angle in each cycle is at the crossing points of the phase voltages. The ideal output voltage, output current, and input current waveforms are shown in Fig. 12(b). The output dc voltage is controlled by varying the firing angle a. The average output voltage under continuous current conduction operation is given by

$$v_o = \frac{3\sqrt{3}}{\pi} E_m \cos\alpha$$

where E_m is the peak value of the phase voltage. At a = 90°, the output voltage is zero. For 0 < a < 90°, vo is positive and power flows from ac supply to the load. For 90° < a < 180°, v_o is negative and the converter operates in the inversion mode. If the load is a dc motor, the power can be transferred from the motor to the ac supply, a process known as regeneration.

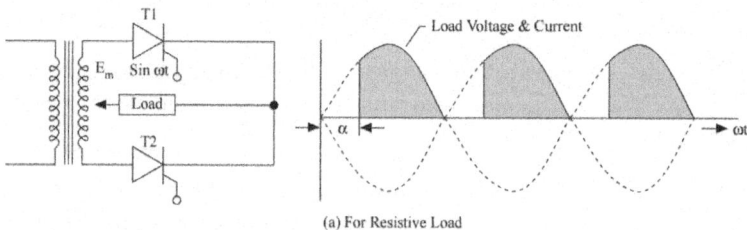

(a) For Resistive Load

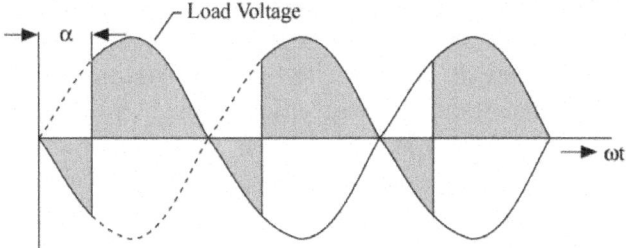

(b) For Resistive-Inductive Load (with continuous current conduction)

Figure 10: Single-phase full-wave converter with transformer.

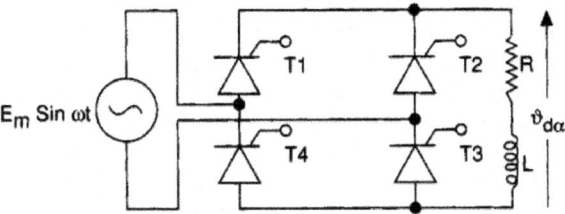

Figure 11: Single-phase bridge converter.

In Fig. 12(a), the top or bottom thyristors could be replaced by diodes. The resulting topology is called a thyristor semiconverter. With this configuration, the input power factor is improved, but the regeneration is not possible.

Cycloconverters

Cycloconverters are direct ac-to-ac frequency changers. The term direct conversion means that the energy does not appear in any form other than the ac input or ac output. The output frequency is lower than the input frequency and is generally an integral multiple of the input frequency. A cycloconverter permits energy to be fed back into the utility network without any additional measures. Also, the phase sequence of the output voltage can be easily reversed by the control system. Cycloconverters have found applications in aircraft systems and industrial drives. These cycloconverters are suitable for synchronous and induction motor control.

DC-to-AC Converters

The dc-to-ac converters are generally called inverters. The ac supply is first converted to dc, which is then converted to a variable-voltage and variable-frequency power supply. This generally consists of a three-phase bridge connected to the ac power source, a dc link with a filter, and the three-phase inverter bridge connectedto the load. In the case of battery-operated systems, there is no intermediate dc link. Inverters can be classified as voltage source inverters (VSIs) and current source inverters (CSIs). A voltage source inverter is fed by a stiff dc voltage, whereas a current source inverter is fed by a stiff current source. A voltage source can be converted to a current source by connecting a series inductance and then varying the voltage to obtain the desired current.

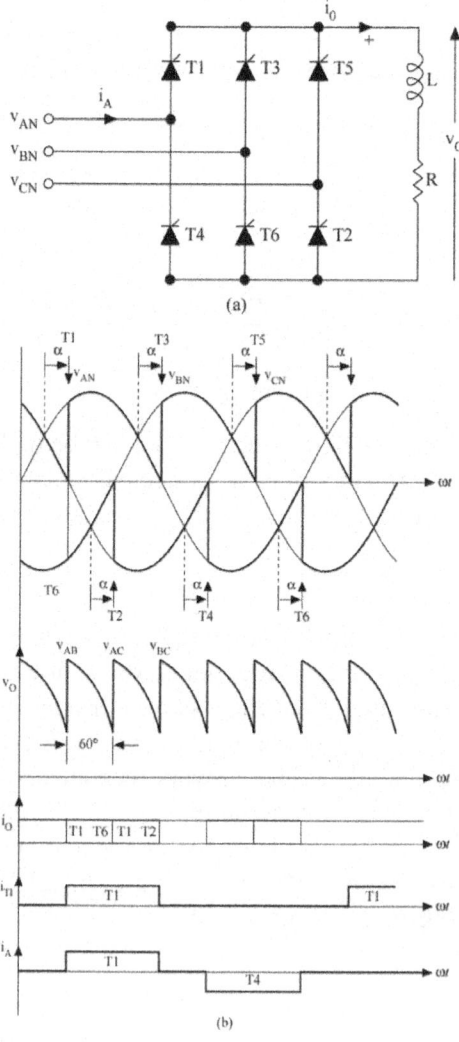

Figure 12: (a) Three-phase thyristor full bridge configuration; (b) output voltage and current waveforms.

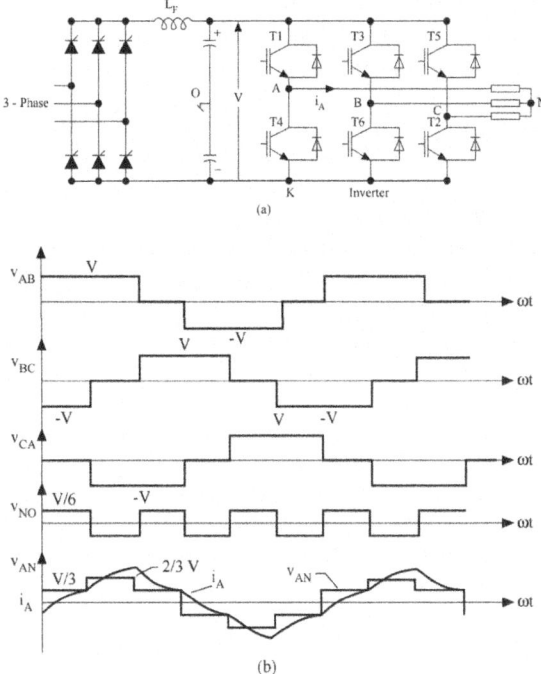

Figure 13: (a) Three-phase converter and voltage source inverter configuration; (b) three-phase square-wave inverter waveforms.

A VSI can also be operated in current-controlled mode, and similarly a CSI can also be operated in the voltagecontrol mode. The inverters are used in variable frequency ac motor drives, uninterrupted power supplies, induction heating, static VAR compensators, etc.

Voltage Source Inverter

A three-phase voltage source inverter configuration is shown in Fig. 13(a). The VSIs are controlled either in square-wave mode or in pulsewidth-modulated (PWM) mode. In square-wave mode, the frequency of the output voltage is controlled within the inverter, the devices being used to switch the output circuit between the plus and minus bus. Each device conducts for 180 degrees, and each of the outputs is displaced 120 degrees to generate a six-step waveform, as shown in Fig. 13(b). The amplitude of the output voltage is controlled by varying the dc link voltage. This is done by varying the firing angle of the thyristors of the three-phase bridge converter at the input. The square-wave-type VSI is not suitable if the dc source is a battery. The six-step output voltage is rich in harmonics and thus needs heavy filtering.

In PWM inverters, the output voltage and frequency are controlled within the inverter by varying the width of the output pulses. Hence at the front end, instead of a phase-controlled thyristor converter, a diode bridge rectifier can be used.A very popular method of controlling the voltage and frequency is by sinusoidal pulsewidth modulation. In this method, a high-frequency triangle carrier wave is compared with a three-phase sinusoidal waveform, as shown in Fig. 14. The power devices in each phase are switched on at the intersection of sineand triangle waves. The amplitude and frequency of the output voltage are varied, respectively, by varying the amplitude and frequency of the reference sine waves. The ratio of the amplitude of the sine wave to the amplitude of the carrier wave is called the modulation index.

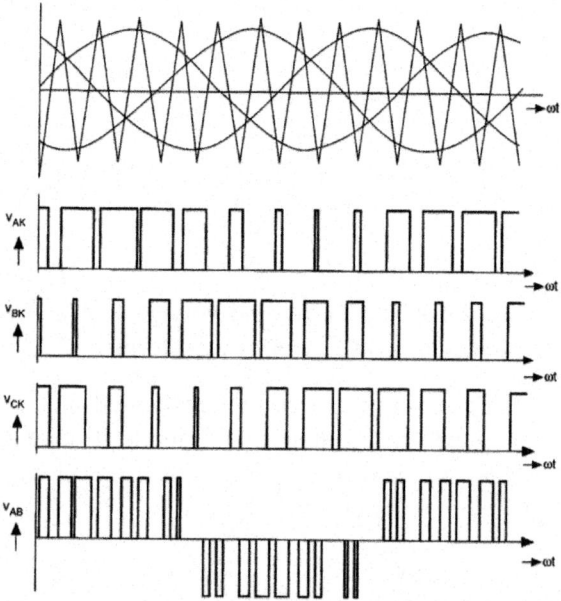

Figure 14: Three-phase sinusoidal PWM inverter waveforms.

The harmonic components in a PWM wave are easily filtered because they are shifted to a higher-frequency region. It is desirable to have a high ratio of carrier frequency to fundamental frequency to reduce the harmonics of lower-frequency components. There are several other PWM techniques mentioned in the literature. The most notable ones are selected harmonic elimination, hysteresis controller, and space vector PWM technique. In inverters, if SCRs are used as power switching devices, an external forced commutation circuit has to be used to turn off the devices. Now, with the availability of IGBTs above 1000-A, 1000-V ratings, they are being used in applications up to 300-

kW motor drives. Above this power rating, GTOs are generally used. Power Darlington transistors, which are available up to 800 A, 1200 V, could also be used for inverter applications.

Current Source Inverter

Contrary to the voltage source inverter where the voltage of the dc link is imposed on the motor windings, in the current source inverter the current is imposed into the motor. Here the amplitude and phase angle of the motor voltage depend on the load conditions of the motor.

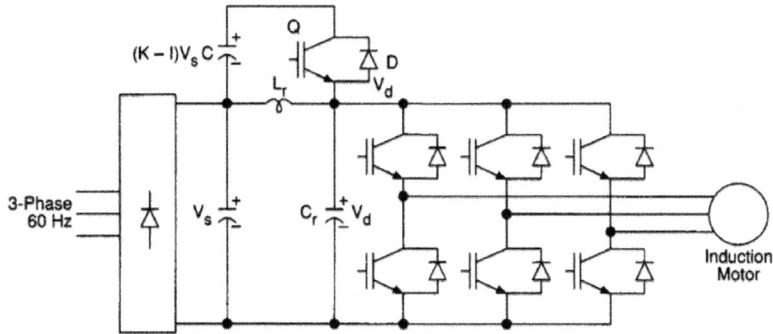

Figure 15: Resonant dc-link inverter system with active voltage clamping.

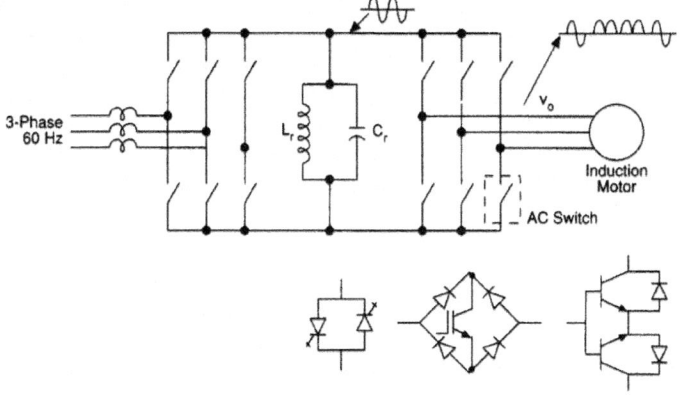

Figure 16: Resonant ac-link converter system showing configuration of ac switches.

Resonant-Link Inverters

The use of resonant switching techniques can be applied to inverter topologies to reduce the switching losses in the power devices. They also permit high switching frequency operation to reduce the size of the magnetic components in the inverter unit. In the resonant dc-link inverter shown in Fig. 15, a resonant circuit is added at the inverter input to convert a fixed dc voltage to a pulsating dc voltage. This resonant circuit enables the devices to be turned on and turned off during the zero voltage interval. Zero voltage or zero current switching is often termed soft switching. Under soft switching, the switching losses in the power devices are almost eliminated. The electromagnetic interference (EMI) problem is less severe because resonant voltage pulses have lower dv/dt compared to those of hard-switched PWM inverters. Also, the machine insulation is less stretched because of lower dv/dt resonant voltage pulses. In Fig. 15, all the inverter devices are turned on simultaneously to initiate a resonant cycle. The commutation from one device to another is initiated at the zero dc-link voltage. The inverter output voltage is formed by the integral numbers of quasi-sinusoidal pulses. The circuit consisting of devices Q, D, and the capacitor C acts as an active clamp to limit the dc voltage to about 1.4 times the diode rectifier voltage V_s.

There are several other topologies of resonant link inverters mentioned in the literature. There are also resonant link ac-ac converters based on bidirectional ac switches, as shown in Fig. 16. These resonant link converters find applications in ac machine control and uninterrupted power supplies, induction heating, etc. The resonant link inverter technology is still in the development stage for industrial applications.

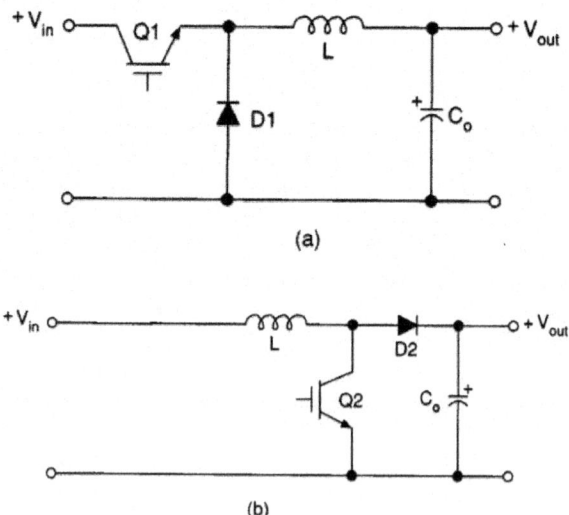

(a)

(b)

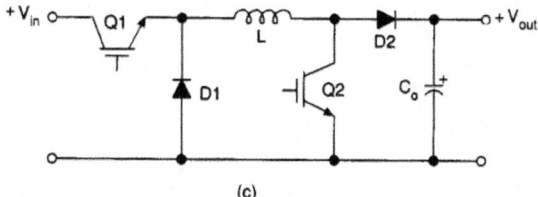

(c)

Figure 17: DC-DC converter configurations: (a) buck converter; (b) boost converter; (c) buck-boost converter.

DC-DC Converters

DC-dc converters are used to convert unregulated dc voltage to regulated or variable dc voltage at the output. They are widely used in switch-mode dc power supplies and in dc motor drive applications. In dc motor control applications, they are called chopper-controlled drives. The input voltage source is usually a battery or derived from an ac power supply using a diode bridge rectifier. These converters are generally either hard-switched PWM types or soft-switched resonant-link types. There are several dc-dc converter topologies, the most common ones being buck converter, boost converter, and buck-boost converter, shown in Fig. 17.

Buck Converter

A buck converter is also called a step-down converter. Its principle of operation is illustrated by referring to Fig. 17(a). The IGBT acts as a high-frequency switch. The IGBT is repetitively closed for a time t_{on} and opened for a time t_{off}. During t_{on}, the supply terminals are connected to the load, and power flows from supply to the load. During t_{off}, load current flows through the freewheeling diode D_1, and the load voltage is ideally zero. The average output voltage is given by

$$V_{out} = DV_{in}$$

where D is the duty cycle of the switch and is given by $D = t_{on}/T$, where T is the time for one period. $1/T$ is the switching frequency of the power device IGBT.

Boost Converter

A boost converter is also called a step-up converter. Its principle of operation is illustrated by referring to Fig. 17(b). This converter is used to produce higher voltage at the load than the supply voltage. When thepower switch is on, the inductor is connected to the dc source and the energy from the supply is stored in it.

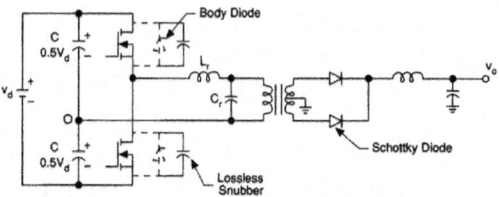

Figure 18: Resonant-link dc-dc converter.

When the device is off, the inductor current is forced to flow through the diode and the load. The induced voltage across the inductor is negative. The inductor adds to the source voltage to force the inductor current into the load. The output voltage is given by

$$V_{out} = \frac{V_{in}}{1 - D}$$

Thus for variation of D in the range $0 < D < 1$, the load voltage Vout will vary in the range $V_{in} < V_{out} < \infty$.

Buck-Boost Converter

A buck-boost converter can be obtained by the cascade connection of the buck and the boost converter. The steady-state output voltage V_{out} is given by

$$V_{out} = V_{in} \frac{D}{1 - D}$$

This allows the output voltage to be higher or lower than the input voltage, based on the duty cycle D. A typical buck-boost converter topology is shown in Fig. 17(c). When the power device is turned on, the input provides energy to the inductor and the diode is reverse biased. When the device is turned off, the energy stored in the inductor is transferred to the output. No energy is supplied by the input during this interval. In dc power supplies, the output capacitor is assumed to be very large, which results in a constant output voltage. In dc drive systems, the chopper is operated in step-down mode during motoring and in step-up mode during regeneration operation.

Resonant-Link DC-DC Converters

The use of resonant converter topologies would help to reduce the switching losses in dc-dc converters and enable the operation at switching frequencies in the megahertz range. By operating at high frequencies, the size of the power supplies could be reduced. There are several types of resonant converter topologies. The most popular configuration is shown in Fig. 18. The dc power is converted to high-frequency alternating power using the MOSFET half-bridge inverter. The resonant capacitor voltage is transformer-coupled, rectified using the two Schottky diodes, and then filtered to get output dc voltage. The output voltage is regulated by control of the inverter switching frequency.

Instead of parallel loading as in Fig. 18, the resonant circuit can be series-loaded; that is, the transformer in the output circuit can be placed in series with the tuned circuit. The series resonant circuit provides the short-circuit limiting feature.

There are other forms of resonant converter topologies mentioned in the literature such as quasi-resonant converters and multiresonant converters. These resonant converter topologies find applications in high-density power supplies.

POWER SUPPLIES

Power supplies are used in many industrial and aerospace applications and also in consumer products. Some of the requirements of power supplies are small size, light weight, low cost, and high power conversion efficiency. In addition to these, some power supplies require the following: electrical isolation between the source and load, low harmonic distortion for the input and output waveforms, and high power factor (PF) if the source is ac voltage. Some special power supplies require controlled direction of power flow. Basically two types of power supplies are required: dc power supplies and ac power supplies. The output of dc power supplies is regulated or controllable dc, whereas the output for ac power supplies is ac. The input to these power supplies can be ac or dc.

DC Power Supplies

In these converters, electrical isolation can only be provided by bulky line frequency transformers. The ac source can be rectified with a diode rectifier to get an uncontrolled dc, and then a dc-to-dc converter can be used to get a controlled dc output. Electrical isolation between the input source and the output load can be provided in the dc-to-dc converter using a high-frequency (HF) transformer. Such HF transformers have small size, light weight, and low cost compared to bulky line frequency transformers. Whether the input source is dc (e.g., battery) or ac, dc-to-dc converters form an important part of dc power supplies, and they are explained in this subsection. DC power supplies can be broadly classified as linear and switching power supplies.

A linear power supply is the oldest and simplest type of power supply. The output voltage is regulated by dropping the extra input voltage across a series transistor (therefore, also referred to as a series regulator). They have very small output ripple, theoretically zero noise, large hold-up time (typically 1–2 ms), and fast response. Linear power supplies have the following disadvantages: very low efficiency, electrical isolation can only be on 60-Hz ac side, larger volume and weight, and, in general, only a single output possible. However, they are still used in very small regulated power supplies and in some special applications (e.g., magnet power supplies). Three terminal linear regulator integrated circuits (ICs) are readily available (e.g., mA7815 has +15-V, 1-A output), are easy to use, and have built-in load short-circuit protection. Switching power supplies use power semiconductor switches in the on and off switching states resulting in high efficiency, small size, and light weight. With the availability of fast switching devices, HF magnetics and capacitors, and high-speed control ICs, switching power supplies have become very popular. They can be further classified as pulsewidth-modulated (PWM) converters and resonant converters, and they are explained below.

Pulsewidth-Modulated Converters

These converters employ square-wave pulse width modulation to achieve voltage regulation. The average output voltage is varied by varying the duty cycle of the power semiconductor switch. The voltage waveform across the switch and at the output are square wave in nature [refer to Fig. 13(b)] and they generally result in higher switching losses when the switching frequency is increased. Also, the switching stresses are high with the generation of large electromagnetic interference (EMI), which is difficult to filter. However, these converters are easy to control, well understood, and have wide load control range.

The methods of control of PWM converters are discussed next.

The Methods of Control

The PWM converters operate with a fixed-frequency, variable duty cycle. Depending on the duty cycle, they can operate in either continuous current mode (CCM) or discontinuous current mode (DCM). If the current through the output inductor never reaches zero (refer to Fig. 13), then the converter operates in CCM; otherwise DCM occurs.

The three possible control methods [Severns and Bloom, 1988; Hnatek, 1981; Unitrode Corporation, 1984; Motorola, 1989; Philips Semiconductors, 1991] are briefly explained below.

- Direct duty cycle control is the simplest control method. A fixed-frequency ramp is compared with the control voltage [Fig. 19(a)] to obtain a variable duty cycle base drive signal for the transistor. This is the simplest method of control. Disadvantages of this method are (a) provides no voltage feedforward to anticipate the effects of input voltage changes, slow response to sudden input changes, poor audio susceptibility, poor open-loop line regulation, requiring higher loop gain to achieve specifications; (b) poor dynamic response.

- Voltage feedforward control. In this case the ramp amplitude varies in direct proportion to the input voltage [Fig. 19(b)]. The open-loop regulation is very good, and the problems in 1(a) above are corrected.

- Current mode control. In this method, a second inner control loop compares the peak inductor current with the control voltage which provides improved open-loop line regulation [Fig. 19(c)]. All the problems of the direct duty cycle control method 1 above are corrected with this method. An additional advantage of this method is that the two-pole second-order filter is reduced to a single-pole (the filter capacitor) first-order filter, resulting in simpler compensation networks.

The above control methods can be used in all the PWM converter configurations explained below.

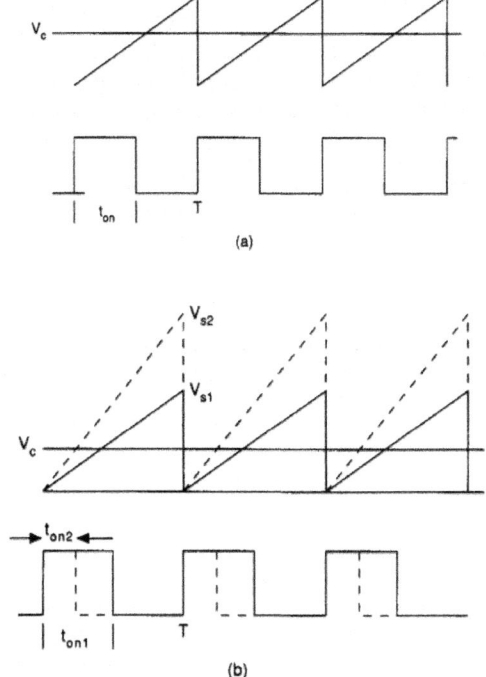

(a)

(b)

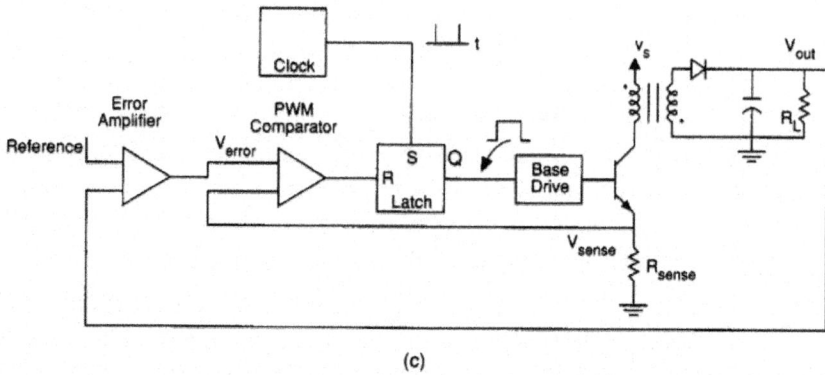

(c)

Figure 19: PWM converter control methods: (a) direct duty cycle control; (b) voltage feedforward control; (c) current mode control (illustrated for flyback converter).

PWM converters can be classified as single-ended and double-ended converters. These converters may or may not have a high-frequency transformer for isolation.

Nonisolated Single-Ended PWM Converters

The basic nonisolated single-ended converters are (a) buck (step-down), (b) boost (step-up), (c) buck-boost (step-up or step-down, also referred to as flyback), and (d) 'Cuk converters (Fig. 20). The 'Cuk converter provides the advantage of nonpulsating input-output current ripple requiring smaller size external filters. Output voltage expression is the same as the buck-boost converter and can be less than or greater than the input voltage. There are many variations of the above basic nonisolated converters, and most of them use a high-frequency transformer for ohmic isolation between the input and the output. Some of them are discussed below.

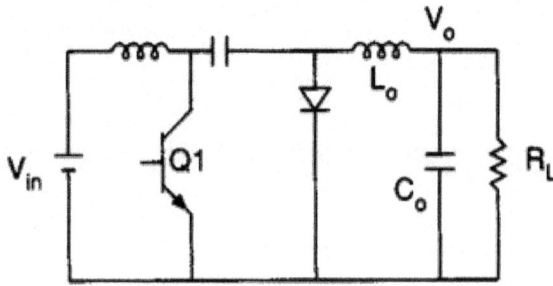

Figure 20: Nonisolated C′ uk converter

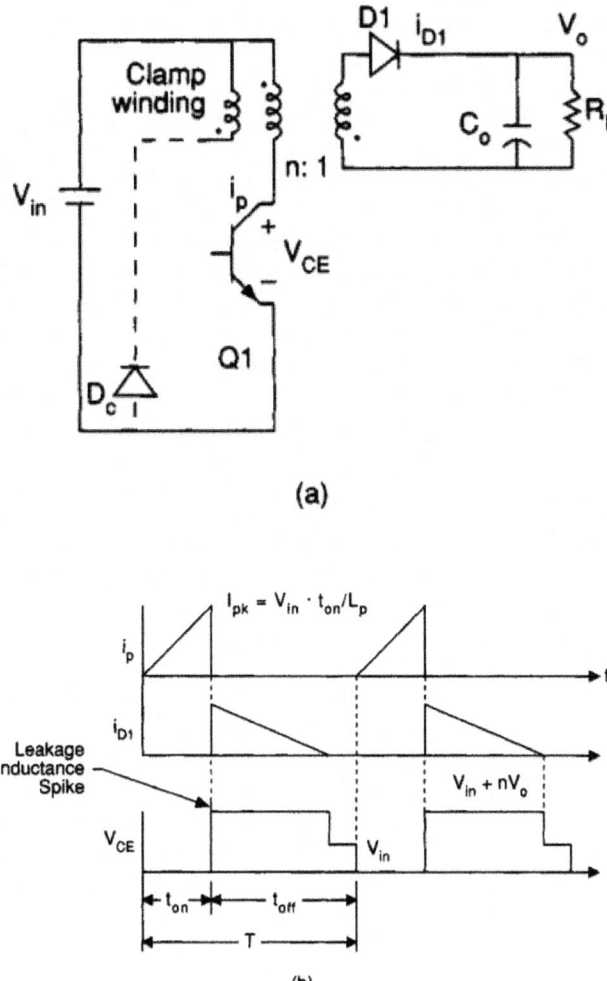

(a)

(b)

Figure 21: (a) Flyback converter. The clamp winding shown is optional and is used to clamp the transistor voltage stress to $V_{in} + nV_o$. (b) Flyback converter waveforms without the clamp winding. The leakage inductance spikes vanish with the clamp winding.

Isolated Single-Ended Topologies

1. The flyback converter (Fig. 21) is an isolated version of the buck-boost converter. In this converter (Fig. 21), when the transistor is on, energy is stored in the coupled inductor (not a transformer), and this energy is transferred to the load when the switch is off. Some of the advantages of this converter are that the leakage inductance is in series

with the output diode when current is delivered to the output, and, therefore, no filter inductor is required; cross regulation for multiple output converters is good; it is ideally suited for high-voltage output applications; and it has the lowest cost. Some of the disadvantages are that large output filter capacitors are required to smooth the pulsating output current; inductor size is large since air gaps are to be provided; and due to stability reasons, flyback converters are usually operated in the DCM, which results in increased losses. To avoid the stability problem, flyback converters are operated with current mode control explained earlier. Flyback converters are used in the power range of 20 to 200 W.

2. The forward converter (Fig. 22) is based on the buck converter. It is usually operated in the CCM to reduce the peak currents and does not have the stability problem of the flyback converter. The HF transformer transfers energy directly to the output with very small stored energy. The output capacitor size and peak current rating are smaller than they are for the flyback. Reset winding is required to remove the stored energy in the transformer. Maximum duty cycle is about 0.45 and limits the control range. This topology is used for power levels up to about 1 kW.

The flyback and forward converters explained above require the rating of power transistors to be much higher than the supply voltage. The two-transistor flyback and forward converters shown in Fig. 23 limit the voltage rating of transistors to the supply voltage.

The Sepic converter shown in Fig. 24 is another isolated single-ended PWM converter.

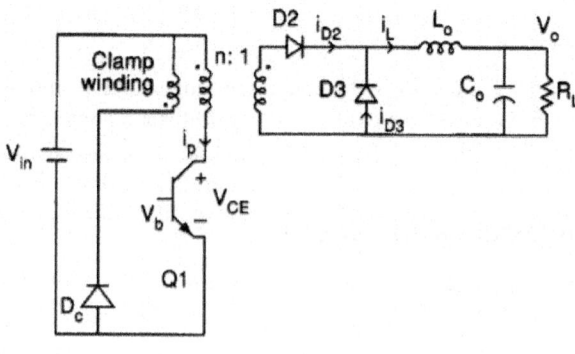

(a)

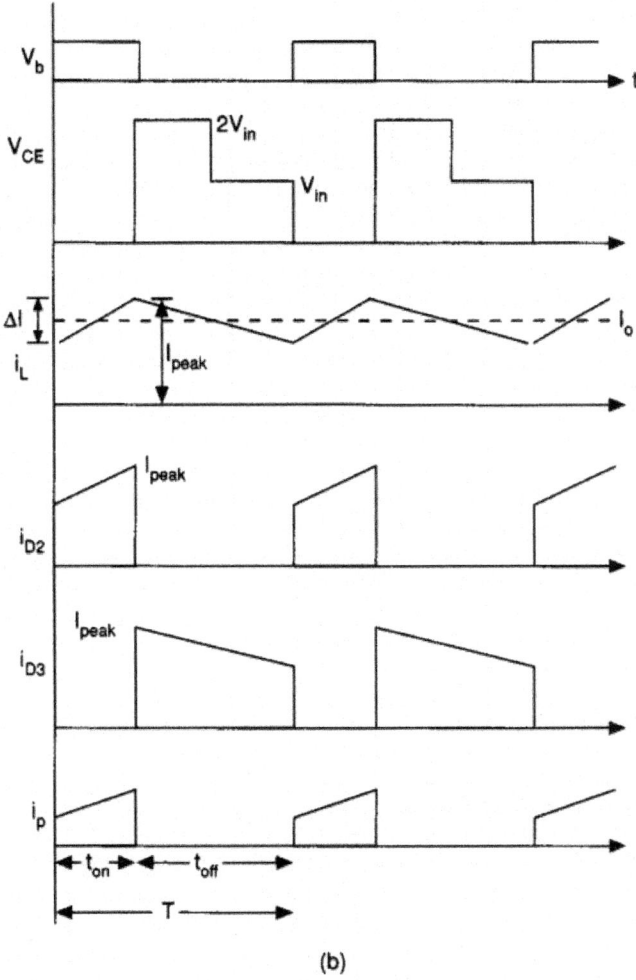

(b)

Figure 22: (a) Forward converter. The clamp winding shown is required for operation. (b) Forward converter waveforms.

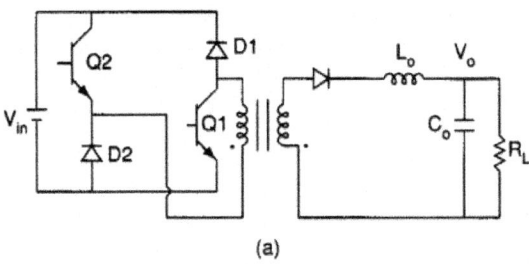

(a)

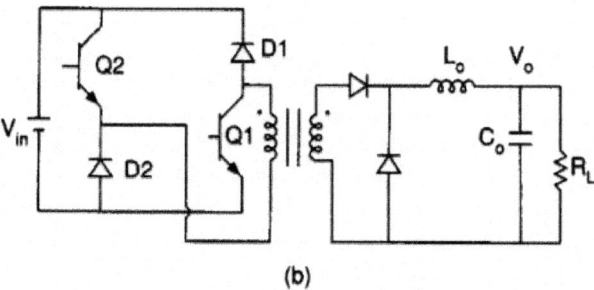

(b)

Figure 23: (a) Two-transistor single-ended flyback converter. (b) Two-transistor single-ended forward converter.

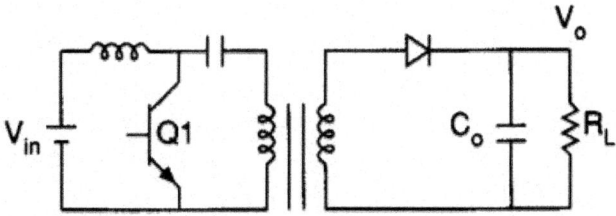

Figure 24: Sepic converter.

Double-Ended PWM Converters

Usually, for power levels above 300 W, double-ended converters are used. In double-ended converters, full-wave rectifiers are used and the output voltage ripple will have twice the switching frequency. Three important double-ended PWM converter configurations are push-pull (Fig. 25), half-bridge (Fig. 26), and full-bridge (Fig. 27).

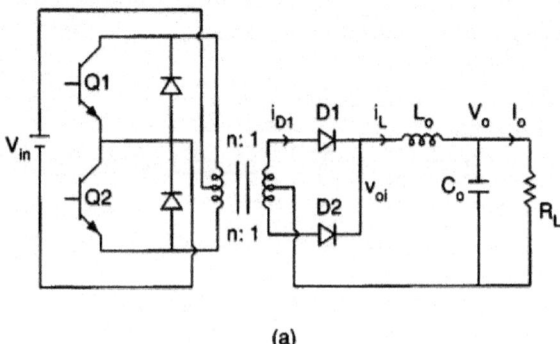

(a)

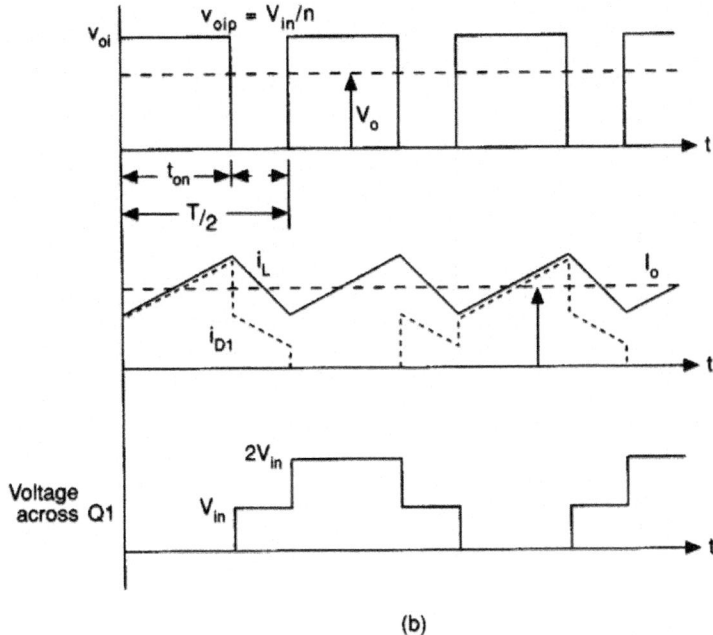

(b)

Figure 25: (a) Push-pull converter and (b) its operating waveforms.

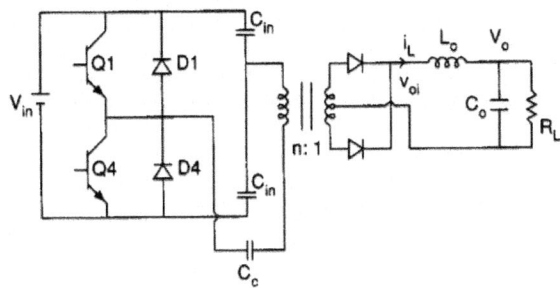

Figure 26: Half-bridge converter. Coupling capacitor C_c is used to avoid transformer saturation.

The Push-Pull Converter: The duty ratio of each transistor in a push-pull converter (Fig. 25) is less than 0.5. Some of the advantages are that the transformer flux swings fully, thereby the size of the transformer is much smaller (typically half the size) than single-ended converters, and output ripple is twice the switching frequency of transistors, allowing smaller filters.

Some of the disadvantages of this configuration are that transistors must block twice the supply voltage, flux symmetry imbalance can cause transformer saturation and special control circuitry is required to avoid this problem, and use of center-tap transformer requires extra copper resulting in higher voltampere (VA) rating.

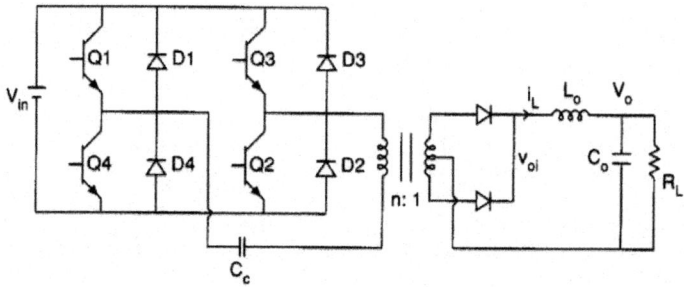

Figure 27: Full-bridge converter.

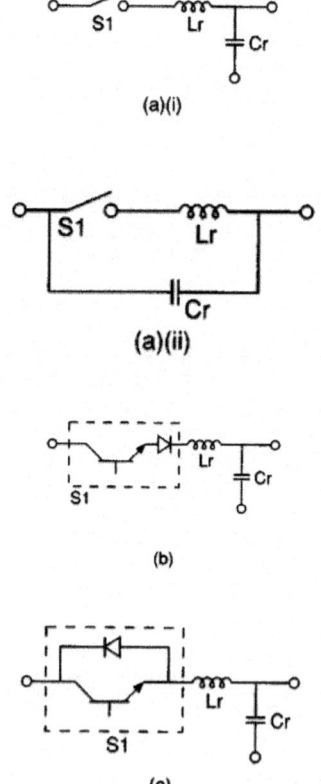

Figure 28: (a) Zero-current resonant switch: (i) L-type and (ii) M-type. (b) Half-wave configuration using L-type ZC resonant switch. (c) Full-wave configuration using L-type ZC resonant switch.

Current mode control (for the primary current) can be used to overcome the flux imbalance. This configuration is used in 100- to 500-W output range.

The Half-Bridge: In the half-bridge configuration (Fig. 26) center-tapped dc source is created by two smoothing capacitors (Cin), and this configuration utilizes the transformer core efficiently. The voltage across each transistor is equal to the supply voltage (half of push-pull) and, therefore, is suitable for high-voltage inputs. One salient feature of this configuration is that the input filter capacitors can be used to change between 110/220-V mains as selectable inputs to the supply. The disadvantage of this configuration is the requirement for large-size input filter capacitors. The half-bridge configuration is used for power levels of the order of 500 to 1000 W.

The full-bridge configuration (Fig. 27) requires only one smoothing capacitor, and for the same transistor type as that of half-bridge, output power can be doubled. It is usually used for power levels above 1 kW, and the design is more costly due to increased number of components (uses four transistors compared to two in push-pull and half-bridge converters). One of the salient features of a full-bridge converter is that by using proper control technique it can be operated in zero-voltage switching (ZVS) mode. This type of operation results in negligible switching losses. However, at reduced load currents, the ZVS property is lost. Recently, there has been a lot of effort to overcome this problem.

Resonant Power Supplies

Similar to the PWM converters, there are two types of resonant converters: single-ended and double-ended. Resonant converter configurations are obtained from the PWM converters explained earlier by adding LC (inductor-capacitor) resonating elements to obtain sinusoidally varying voltage and/or current waveforms. This approach reduces the switching losses and the switch stresses during switching instants, enabling the converter to operate at high switching frequencies,resulting in reduced size, weight, and cost. Some other advantages of resonant converters are that leakage inductances of HF transformers and the junction capacitances of semiconductors canbe used profitably in the resonant circuit, and reduced EMI. The major disadvantage of resonant converters is increased peak current (or voltage) stress.

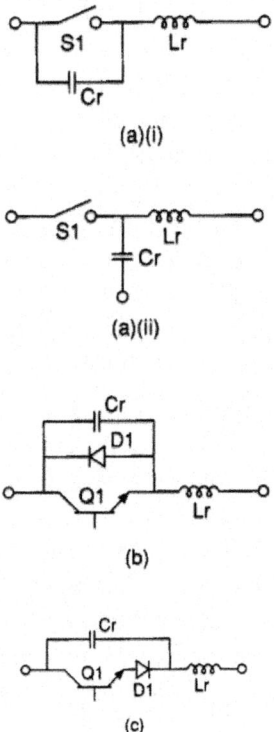

Figure 29: (a) Zero-voltage resonant switches. (b) Half-wave configuration using ZV resonant switch shown in Fig. (a)(i). (c) Full-wave configuration using ZV resonant switch shown in Fig. (a)(i).

Single-Ended Resonant Converters

They are referred to as quasi-resonant converters (QRCs) since the voltage (or current) waveforms are quasi-sinusoidal in nature. The QRCs can operate with zero-current switching (ZCS) or ZVS or both. All the QRC configurations can be generated by replacing the conventional switches by the resonant switches shown in Figs. 28 and 29.A number of configurations are realizable. Basic principles of ZCS and ZVS are explained briefly below.

- Zero-current switching QRCs [Sum, 1988; Liu et al., 1985]. Figure 30(a) shows an example of a ZCS QR buck converter implemented using a ZC resonant switch. Depending on whether the resonant switch is half-wave type or full-wave type, the resonating current will be only half-wave sinusoidal [Fig. 30(b)] or a full sine-wave [Fig. 30(c)]. The device currents are shaped sinusoidally, and, therefore, the switching losses are almost negligible with low turn-on and turn-off stresses. ZCS QRCs can operate at frequencies of the order of 2 MHz. The major problems with this type of converter are high peak currents through the switch and capacitive turn-on losses.

• Zero-voltage switching QRCs [Sum, 1988; Liu and Lee, 1986]. ZVS QRCs are duals of ZCS QRCs. The auxiliary LC elements are used to shape the switching device's voltage waveform at off time in order to create a zero-voltage condition for the device to turn on. Fig. 31(a) shows an example of ZVS QR boost converter implemented using a ZV resonant switch. The circuit can operate in the half-wave mode [Fig. 31(b)] or in the full-wave mode [Fig. 31(c)] depending on whether a half-wave or full-wave ZV resonant switch is used, and the name comes from the capacitor voltage waveform. The full-wave mode ZVS circuit suffers from capacitive turn-on losses. The ZVS QRCs suffer from increased voltage stress on the switch. However, they can be operated at much higher frequencies compared to ZCS QRCs.

Double-Ended Resonant Converters

These converters [Sum, 1988; Bhat, 1991; Steigerwald, 1988; Bhat, 1992] use full-wave rectifiers at the output, and they are generally referred to as resonant converters. A number of resonant converter configurations are realizable by using different resonant tank circuits, and the three most popular configurations, namely, the series resonant converter (SRC), the parallel resonant converter (PRC), and the series-parallel resonant converter (SPRC) (also called LCC-type PRC), are shown in Fig. 32.

Series resonant converters [Fig. 32(a)] have high efficiency from full load to part load. Transformer saturation is avoided due to the series blocking resonating capacitor. The major problems with the SRC are that it requires a very wide change in switching frequency to regulate the load voltage and the output filter capacitor must carry high ripple current (a major problem especially in low output voltage, high output current applications).

Parallel resonant converters [Fig. 32(b)] are suitable for low output voltage, high output current applications due to the use of filter inductance at the output with low ripple current requirements for the filter capacitor. The major disadvantage of the PRC is that the device currents do not decrease with the load current, resulting in reduced efficiency at reduced load currents.

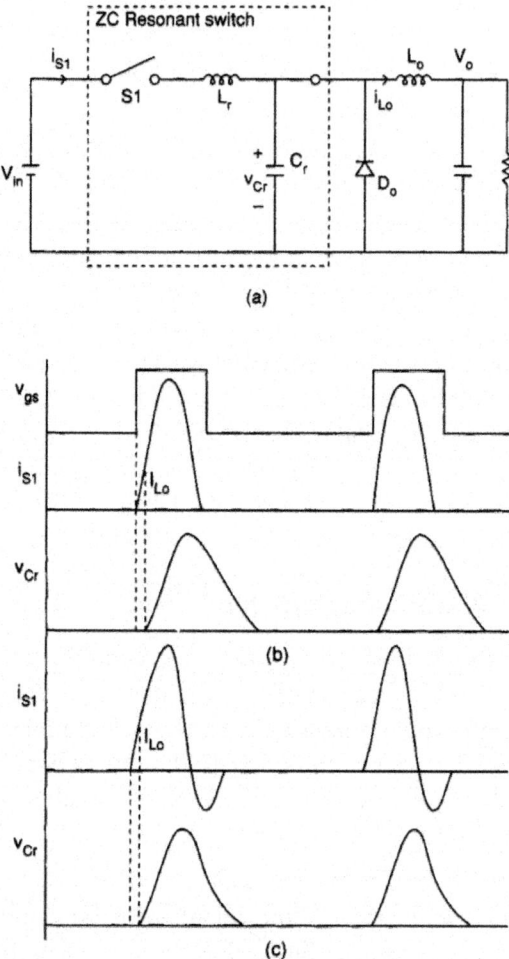

Figure 30: (a) Implementation of ZCS QR buck converter using L-type resonant switch. (b) Operating waveforms for half-wave mode. (c) Operating waveforms for full-wave mode.

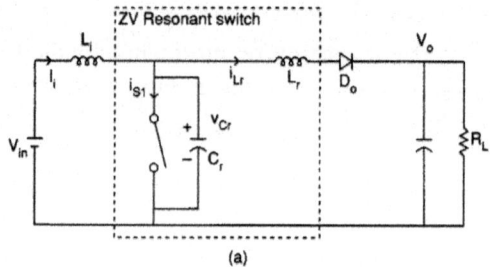

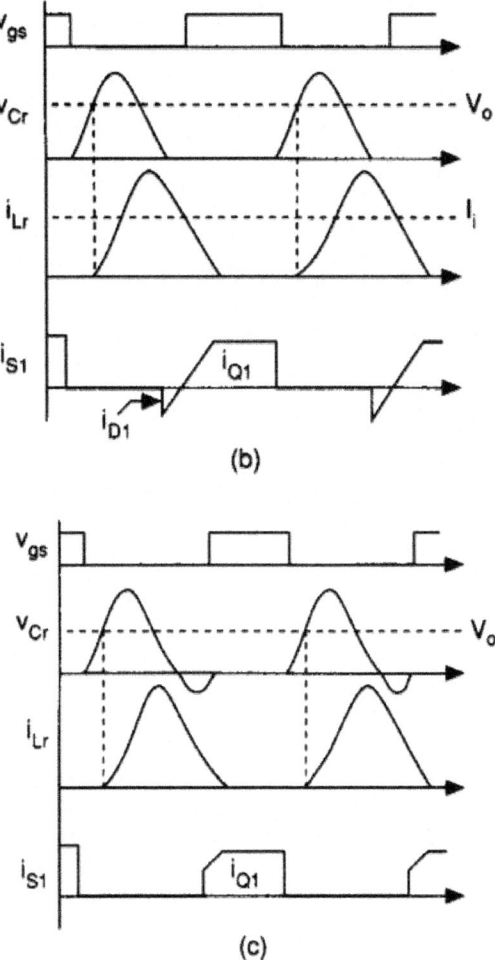

Figure 31: (a) Implementation of ZVS QR buck converter using resonant switch shown in Fig. 28(a)(i). (b) Operating waveforms for half-wave mode. (c) Operating waveforms for full-wave mode.

The SPRC [Fig. 32(c)] takes the desirable features of SRC and PRC. Load voltage regulation in resonant converters for input supply variations and load changes is achieved by either varying the switching frequency or using fixed-frequency (variable pulsewidth) control.

Variable-Frequency Operation: Depending on whether the switching frequency is below or above the natural resonance frequency (w_r), the converter can operate in different operating modes as explained below.

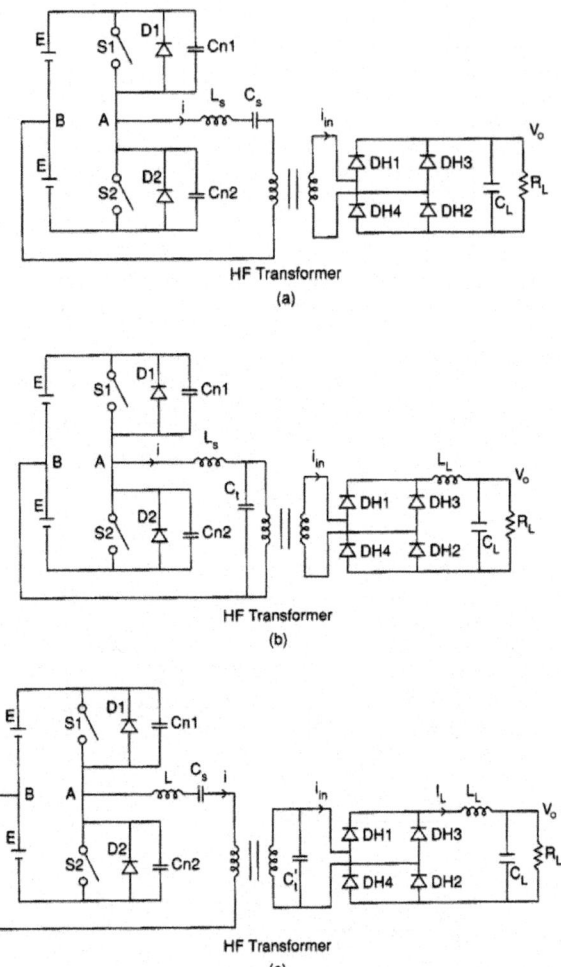

Figure 32: High-frequency resonant converter (half-bridge version) configurations suitable for operation above resonance. Cn1 and Cn2 are the snubber capacitors. (Note: For operation below resonance, di/dt limiting inductors and RC snubbers are required. For operation above resonance, only capacitive snubbers are required as shown.) (a) Series resonant converter. Leakage inductances of the HF transformer can be part of resonant inductance. (b) Parallel resonant converter. (c) Series-parallel (or LCC-type) resonant converter with capacitor C_t placed on the secondary side of the HF transformer.

Below-Resonance (Leading PF) Mode: When the switching frequency is below the natural resonance frequency, the converter operates in a below-resonance mode (Fig. 33). The equivalent impedance across AB presents a leading PF so that natural turn-off of the switches is assured and any type of fast turn-off switch (including asymmetric SCRs) can be used. Depending on the instant of turn-on of switches S_1 and S_2, the converter can enter into two modes of operation, namely, continuous and discontinuous current modes. The steady-state operation in continuous current mode (CCM) [Fig. 33(a)] is explained briefly as follows.

Assume that diode D_2 was conducting and switch S_1 is turned on. The current carried by D_2 will be transferred to S_1 almost instantaneously (except for a small time of recovery of D_2 during which input supply is shorted through D_2 and S_1, and the current is limited by the di/dt limiting inductors). The current i then oscillates sinusoidally and goes to zero in the natural way.

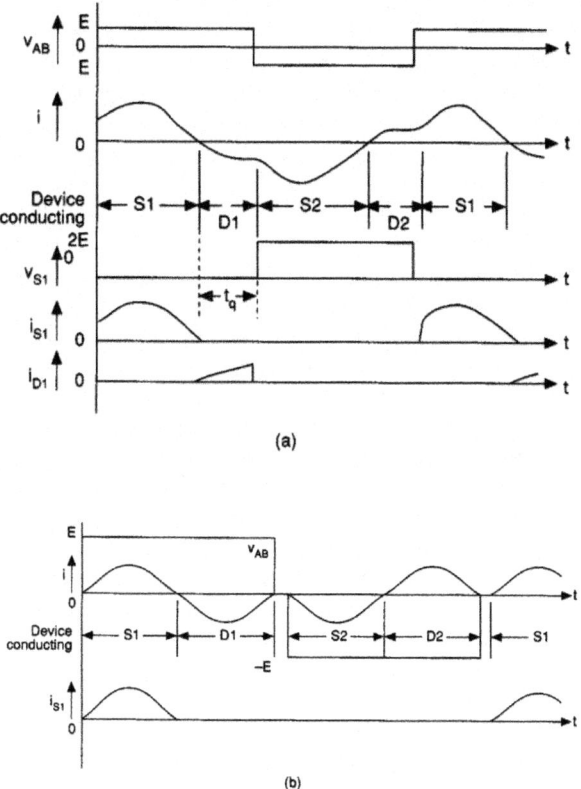

(a)

(b)

Figure 33: Typical waveforms at different points of a resonant converter operating below resonance (a) in continuous current mode and (b) in discontinuous current mode.

The current tries to reverse, and the path for this current is provided by the diode D_1. Conduction of D_1 feeds the reactive energy in the load and the tank circuit back to the supply. The on-state of D_1 also provides a reverse voltage across S_1, allowing it to turn off. After providing a time equal to or greater than the turn-off time of S_1, switch S_2 can be turned on to initiate the second half cycle. The process is similar to the first half cycle, with the voltage across v_{AB} being of opposite polarity, and the functions of D_1, S_1 will be assumed by D_2, S_2. With this type of operation, the converter works in the continuous current mode as the switches are turned on before the currents in the diodes reach zero. If the switching on of S_1 and S_2 is delayed such that the currents through the previously conducting diodes reach zero, then there are zero current intervals and the inverter operates in the DCM [Fig. 33(b)].

Load voltage regulation is achieved by decreasing the switching frequency below the rated value. Since the inverter output current i leads the inverter output voltage v_{AB}, this type of operation is also called a leading PF mode of operation. If transistors are used as the switching devices, then for operation in DCM, the pulsewidth can be kept constant while decreasing the switching frequency to avoid CCM operation. DCM operation has the advantages of negligible switching losses due to ZCS, lower di/dt and dv/dt stresses, and simple control circuitry. However, DCM operation results in higher switch peak currents. From the waveforms shown in Fig. 33, the following problems can be identified for operation in the below-resonance mode: requirement of di/dt inductors to limit the large turn-on switch currents and a need for lossy RC snubbers and fast recovery diodes. Since the switching frequency is decreased to control the load power, the HF transformer and magnetics must be designed for the lowest switching frequency, resulting in increased size of the converter.

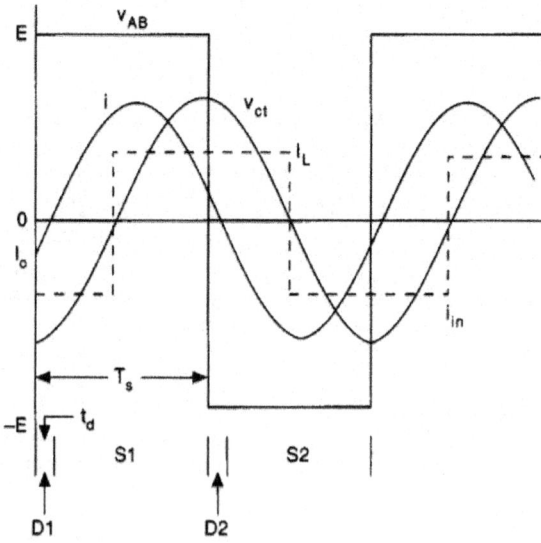

Figure 34: Typical operating waveforms at different points of an SPRC operating above resonance.

Above-Resonance (Lagging PF) Mode

If switches capable of gate or base turn-off (e.g., MOSFETs, bipolar transistors) are used, then the converter can operate in the above-resonance mode (lagging PF mode). Figure 34 shows some typical operating waveforms for such type of operation, and it can be noticed that the current i lags the voltage v_{AB}. Since the switch takes current from its own diode across it at zero-current point, there is no need for di/dt limiting inductance, and a simple capacitive snubber can be used. In addition, the internal diodes of MOSFETs can be used due to the large turn-off time available for the diodes. Major problems with the lagging PF mode of operation are that there are switch turn-off losses, and since the voltage regulation is achieved by increasing the switching frequency above the rated value, the magnetic losses increase and the design of a control circuit is difficult.

Exact analysis of resonant converters is complex due to the nonlinear loading on the resonant tanks. The rectifier-filter-load resistor block can be replaced by a square-wave voltage source [for SRC, Fig. 32(a)] or a square-wave current source [for PRC and SPRC, Fig. 32(b) and (c)]. Using fundamental components of the waveforms, an approximate analysis [Bhat, 1991; Steigerwald, 1988] using a phasor circuit gives a reasonably good design approach. This analysis approach is illustrated next for the SPRC.

Approximate Analysis of SPRC

Figure 35 shows the equivalent circuit at the output of the inverter and the phasor circuit used for the analysis. All the equations are normalized using the base quantities

Base voltage $V_B = E_{min}$

Base impedance $Z_B = R'_L = n^2 R_L$

Base current $I_B = V_B/I_B$

The converter gain [normalized output voltage in per unit (p.u.) referred to the primary-side] can be derived as [Bhat, 1991; Steigerwald, 1988]

$$M = \cfrac{1}{\left\{ \left(\cfrac{\pi^2}{8} \right)^2 \left[1 + \left(\cfrac{C_t}{C_s} \right) \left(1 - y_s^2 \right) \right]^2 + Q_s^2 \left[y_s - \left(\cfrac{1}{y_s} \right) \right]^2 \right\}^{1/2}} \quad \text{p.u.}$$

(1)

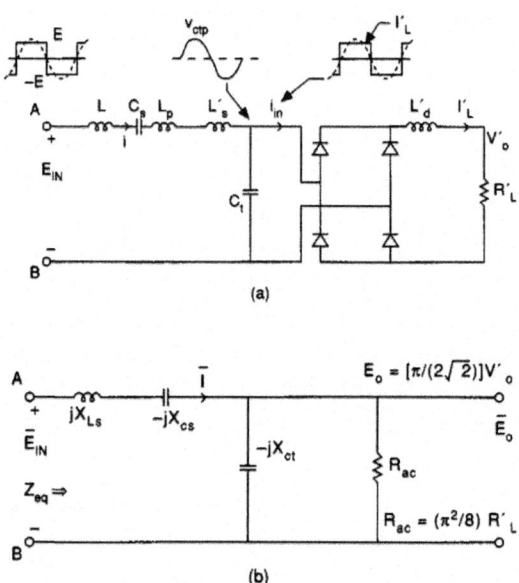

Figure 35: (a) Equivalent circuit for a SPRC at the output of the inverter terminals (across AB) of Fig. 31(c), Lp and L¢s are the leakage inductance of the primary and primary referred leakage inductance of the secondary, respectively. (b) Phasor circuit model used for the analysis of the SPRC converter.

$$Q_s = \frac{(L_s / C_s)^{1/2}}{R'_L} \; ; \; L_s = L + L_p + L'_s \tag{2}$$

$$y_s = \frac{f_s}{f_r} \tag{3}$$

f_s = switching frequency

f_r = series resonance frequency

$$= \frac{\omega_r}{2\pi} = \frac{1}{2\pi(L_sC_s)^{1/2}} \tag{4}$$

The equivalent impedance looking into the terminals AB is given by

$$Z_{eq} = \frac{B_1 + jB_2}{B_3} \; \text{p.u.} \tag{5}$$

where

$$B_1 = \left(\frac{8}{\pi^2}\right)\left(\frac{C_s}{C_t}\right)^2\left(\frac{Q_s}{y_s}\right)^2$$

(6)

$$B_2 = Q_s\left(y_s - \frac{1}{y_s}\right)\left[1 + \left(\frac{8}{\pi^2}\right)^2\left(\frac{C_s}{C_t}\right)^2\left(\frac{Q_s}{y_s}\right)^2\right] - \left(\frac{C_s}{C_t}\right)\left(\frac{Q_s}{y_s}\right)$$

(7)

$$B_3 = 1 + \left(\frac{8}{\pi^2}\right)^2\left(\frac{C_s}{C_t}\right)^2\left(\frac{Q_s}{y_s}\right)^2$$

(8)

The peak inverter output (resonant inductor) current can be calculated using

$$I_p = \frac{4}{\pi|Z_{eq}|} \quad \text{p.u.}$$

(9)

The same current flows through the switching devices.

The value of initial current I_0 is given by

$$I_0 = I_p \sin(-\phi) \quad \text{p.u.}$$

(10)

where $\phi = \tan^{-1}(B_2/B_1)$ rad. B_1 and B_2 are given by Eqs. (6) and (7), respectively.

If I_0 is negative, then forced commutation is necessary and the converter is operating in the lagging PF mode. The peak voltage across the capacitor C'_t (on the secondary side) is

$$V_{ctp} = \frac{\pi}{2} V_0 \quad \text{V}$$

(11)

The peak voltage across C_s and the peak current through C'_t are given by

$$V_{csp} = \frac{Q_s}{y_s} I_p \quad \text{p.u.}$$

(12)

$$I_{ctp} = \frac{V_{ctp}}{X_{cptu} R_L} \quad A \tag{13}$$

$$X_{ctpu} = \left(\frac{C_s}{C_t}\right)\left(\frac{Q_s}{y_s}\right) \quad \text{p.u.} \tag{14}$$

The plot of converter gain versus the switching frequency ratio y_s, obtained using (1), is shown for $C_s / C_t = 1$ in Fig. 36, for the lagging PF mode of operation. If the ratio C_s / C_t increases, then the converter takes the characteristics of SRC and the load voltage regulation requires a very wide range in the frequency change. Lower values of C_s / C_t take the characteristics of a PRC. Therefore, a compromised value of $C_s / C_t = 1$ is chosen.

It is possible to realize higher-order resonant converters with improved characteristics and many of them are presented in Bhat [1991].

Fixed-Frequency Operation

To overcome some of the problems associated with the variable frequency control of resonant converters, they are operated with fixed frequency [Sum, 1988; Bhat, 1992].A number of configurations and control methods for fixed-frequency operation are available in the literature (Bhat [1992] gives a list of papers). One of the most popular methods of control is the phase-shift control(also called clamped-mode or PWM operation) method. Figure 37 illustrates the clamped-mode fixed-frequency operation of the SPRC. The load power control is achieved by changing the phase-shift angle ϕ between the gating signals to vary the pulsewidth of v_{AB}.

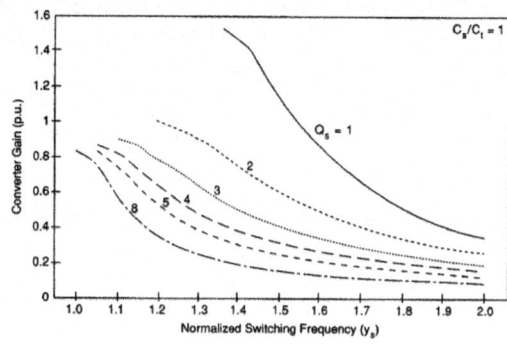

Figure 36: The converter gain M (p.u.) (normalized output voltage) versus normalized switching frequency y_s of SPRC operating above resonance for $C_s/C_t = 1$.

Design Example

Design a 500-W output SPRC (half-bridge version) with secondary-side resonance (operation in lagging PF mode and variable-frequency control) with the following specifications:

$$\text{Minimum input supply voltage} = 2E_{min} = 230 \text{ V}$$

$$\text{Load voltage, } V_o = 48 \text{ V}$$

$$\text{Switching frequency, } f_s = 100 \text{ kHz}$$

$$\text{Maximum load current} = 10.42 \text{ A}$$

As explained in item 2, $C_s/C_t = 1$ is chosen. Using the constraints (1) minimum kVA rating of tank circuit per kW output power, (2) minimum inverter output peak current, and (3) enough turn-off time for the switches, it can be shown that [Bhat, 1991] $Q_s = 4$ and $y_s = 1.1$ satisfy the design constraints. From Fig. 36, M = 0.8 p.u.

Average load voltage referred to the primary side of the HF transformer = 0.8 2 115 V = 92 V. Therefore, the transformer turns ratio required .1.84.

$$R'_L = n^2 \left(\frac{V_o^2}{P_o} \right) = 15.6 \ \Omega$$

The values of L_s and C_s can be obtained by solving

$$\left(\frac{L_s}{C_s} \right)^{1/2} = 4 \times 15.6 \ \Omega \text{ and } \omega_r = \frac{1}{(L_s C_s)^{1/2}} = 2\pi \frac{f_s}{y_s}$$

Solving the above equations gives $L_s = 109$ mH and $C_s = 0.0281$ mF. Leakage inductance ($L_p + L'_s$) of the HF transformer can be used as part of L_s. Typical value for a 100-kHz practical transformer (using TokinMn-Zn 2500B2 Ferrite, E-I type core) for this application is about 5 mH. Therefore, the external resonant inductance required is L = 104 mH.

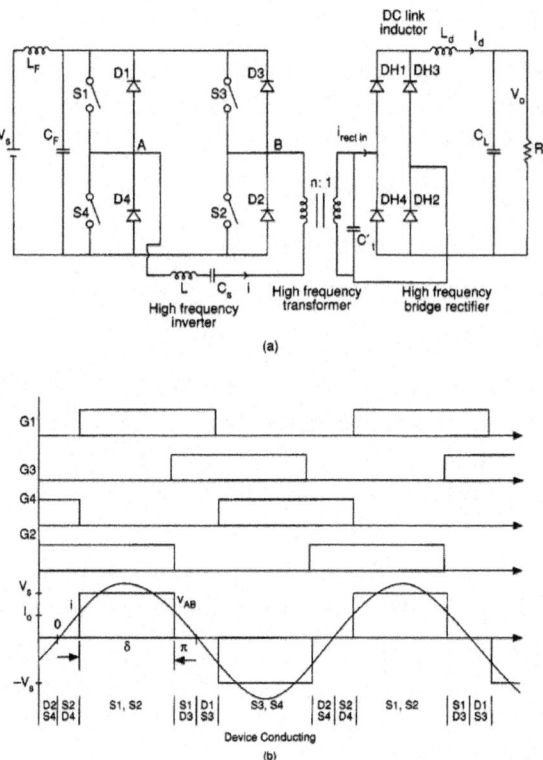

Figure 37: (a) Basic circuit diagram of series-parallel resonant converter suitable for fixed-frequency operation with PWM (clamped-mode) control. (b) Waveforms to illustrate the operation of fixed-frequency PWM series-parallel resonant converter working with a pulsewidth d.

Since $C_s/C_t = 1$ is chosen, $C_t = 0.0281$ mF. The actual value of Ct used on the secondary side of the HF transformer $= (1.84)^2$ 2 $0.0281 = 0.09514$ mF. The resonating capacitors must be HF type (e.g., polypropylene) and must be capable of withstanding the voltage and current ratings obtained above (enough safety margin must be provided).

Using Eqs. (9) and (11) to (13):

$$\text{Peak current through switches} = 7.6 \text{ A}$$

$$\text{Peak voltage across } C_s, V_{csp} = 430 \text{ V}$$

$$\text{Peak voltage (on secondary side) across } C'_t, V_{ctp} = 76 \text{ V}$$

$$\text{Peak current through capacitor } C'_t \text{ (on secondary side), } I_{ctp} = 4.54 \text{ A}$$

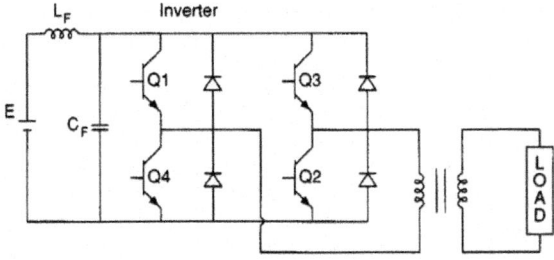

Figure 38: An inverter circuit to obtain variable-voltage, variable-frequency ac source. Using sinusoidal pulse width modulation control scheme, sine-wave ac output voltage can be obtained.

A simple control circuit can be built using PWM IC SG3525 and TSC429 MOSFET driver ICs. With the development of digital ICs operating on low-voltage (of the order of 3 V) supplies, use of MOSFETs as synchronous rectifiers with very low voltage drop (~0.2 V) has become essential [Motorola, 1989] to increase the efficiency of the power supply.

AC Power Supplies

Some applications of ac power supplies are ac motor drives, uninterruptible power supply (UPS) used as a standby ac source for critical loads (e.g., in hospitals, computers), and dc sourceto-utility interface (either to meet peak power demands or to augment energy by connecting unconventional energy sources like photovoltaic arrays to the utility line). In ac induction motor drives, the ac power main is rectified and filtered to obtain a smooth dc source, and then an inverter (single-phase version is shown in Fig. 38) is used to obtain a variable-frequency, variable-voltage ac source. Some other methods used to get sinusoidal voltage output are [Rashid, 1988] a number of phase-shifted inverter outputs summed in an output transformer to get a stepped waveform that approximates a sine wave and the use of a bang-bang controller in Fig. 38. All these methods use linefrequency (60 Hz) transformers for voltage translation and isolation purposes. To reduce the size, weight, and cost of such systems, one can use dc-to-dc converters (discussed earlier) as an intermediate stage. Figure 39 shows such a system in block schematic form. One can use an HF inverter circuit (discussed earlier) followed by a cycloconverter stage. The major problem with these schemes is the reduction in efficiency due to the extra power stage. Figure 40 shows a typical UPS scheme. The battery shown has to be charged by a separate rectifier circuit. AC-to-ac conversion can also be achieved using cycloconverters [e.g., Rashid, 1988].

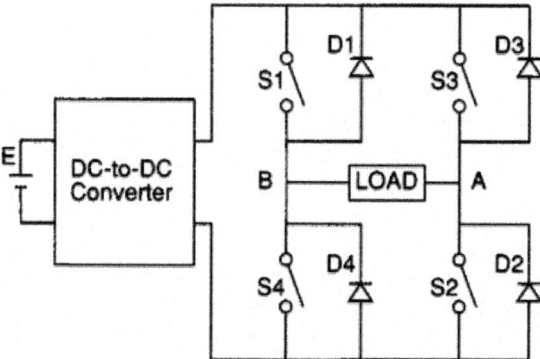

Figure 39: AC power supplies using HF switching (PWM or resonant) dc-to-dc converter as an input stage. HF transformer isolated dc-todc converters can be used to reduce the size and weight of the power supply. Sinusoidal voltage output can be obtained using the modulation in the output inverter stage or in the dc-to-dc converter.

Special Power Supplies

Using the inverters and cycloconverters, it is possible to realize bidirectional ac and dc power supplies. In these power supplies [Rashid, 1988], power can flow in both directions, i.e., from input to output or from output to input. It is also possible to control the ac-to-dc converters to obtain sinusoidal line current with unity PF and low harmonic distortion at the ac source.

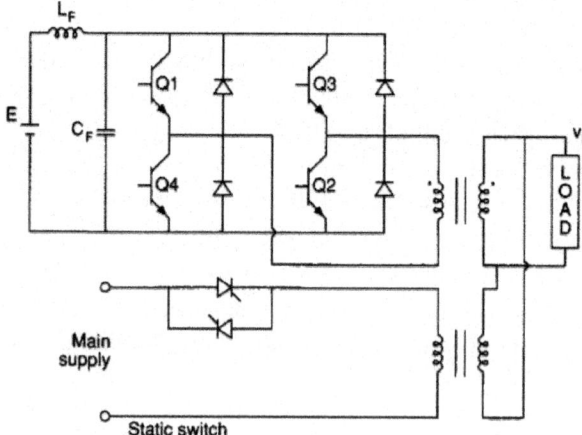

Figure 40: A typical arrangement of UPS system. The load gets power through the static switch when the ac main supply is present. The inverter supplies power when the main supply fails.

CONVERTER CONTROL OF MACHINES

Converter-controlled electrical machine drives are very important in modern industrial applications. Some examples in the high-power range are metal rolling mills, cement mills, and gas line compressors. In the medium-power range are textile mills, paper mills, and subway car propulsion. Machine tools and computer peripherals are examples of converter-controlled electrical machine drive applications in the low-power range. The converter normally provides a variable-voltage dc power source for a dc motor drive and a variablefrequency, variable-voltage ac power source for an ac motor drive. The drive system efficiency is high because the converter operates in switching mode using power semiconductor devices. The primary control variable of the machine may be torque, speed, or position, or the converter can operate as a solid-state starter of the machine. The recent evolution of high-frequency power semiconductor devices and high-density and economical microelectronic chips, coupled with converter and control technology developments, is providing a tremendous boost in the applications of drives.

Converter Control of DC Machines

The speed of a dc motor can be controlled by controlling the dc voltage across its armature terminals. A phasecontrolled thyristor converter can provide this dc voltage source. For a low-power drive, a single-phase bridge converter can be used, whereas for a high-power drive, a three-phase bridge circuit is preferred. The machine can be a permanent magnet or wound field type. The wound field type permits variation and reversal of field and is normally preferred in large power machines.

Phase-Controlled Converter DC Drive

Figure 41 shows a dc drive using a three-phase thyristor bridge converter. The converter rectifies line ac voltage to variable dc output voltage by controlling the firing angle of the thyristors. With rated field excitation, as the armature voltage is increased, the machine will develop speed in the forward direction until the rated, or base, speed is developed at full voltage when the firing angle is zero. The motor speed can be increased further by weakening the field excitation. Below the base speed, the machine is said to operate in constant torque region, whereas the field weakening mode is defined as the constant power region.

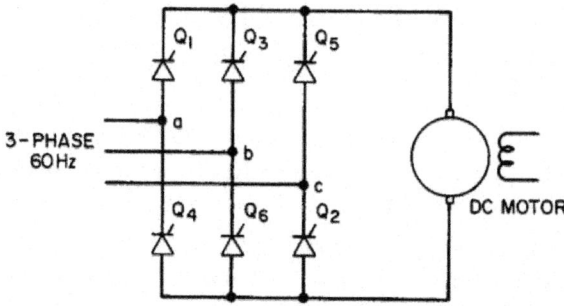

Figure 41: Three-phase thyristor bridge converter control of a dc machine.

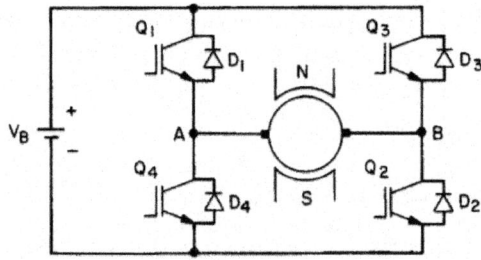

Figure 42: Four-quadrant dc motor drive using an H-bridge converter.

At any operating speed, the field can be reversed and the converter firing angle can be controlled beyond 90 degrees for regenerative braking mode operation of the drive. In this mode, the motor acts as a generator (with negative induced voltage) and the converter acts as an inverter so that the mechanical energy stored in the inertia is converted to electrical energy and pumped back to the source. Such two-quadrant operation gives improved efficiency if the drive accelerates and decelerates frequently. The speed of the machine can be controlled with precision by a feedback loop where the command speed is compared with the machine speed measured by a tachometer. The speed loop error generally generates the armature current command through a compensator. The current is then feedback controlled with the firing angle control in the inner loop. Since torque is proportional to armature current (with fixed field), a current loop provides direct torque control, and the drive can accelerate or decelerate with the rated torque. A second bridge converter can be connected in antiparallel so that the dual converter can control the machine speed in all the four quadrants (motoring and regeneration in forward and reverse speeds).

Pulsewidth Modulation Converter DC Machine Drive

Four-quadrant speed control of a dc drive is also possible using an H-bridge pulsewidth modulation (PWM) converter as shown in Fig. 42. Such drives (using a permanent magnet dc motor) are popular in low-power applications, such as robotic and instrumentation drives. The dc source can be a battery or may be obtained from ac supply through a diode rectifier and filter. With PWM operation, the drive response is very fast and the armature current ripple is small, giving less harmonic heating and torque pulsation. Four-quadrant operation can be summarized as follows:

Quadrant 1: Forward motoring (buck or step-down converter mode)

Q1—on

Q3, Q4—off

Q2—chopping

Current freewheeling through D3 and Q1

Quadrant 2: Forward regeneration (boost or step-up converter mode)

Q1, Q2, Q3—off

Q4—chopping

Current freewheeling through D1 and D2

Quadrant 3: Reverse motoring (buck converter mode)

Q3—on

Q1, Q2—off

Q4—chopping

Current freewheeling through D1 and Q3

Quadrant 4: Reverse regeneration (boost converter mode)

Q1, Q3, Q4—off

Q2—chopping

Current freewheeling through D3 and D4

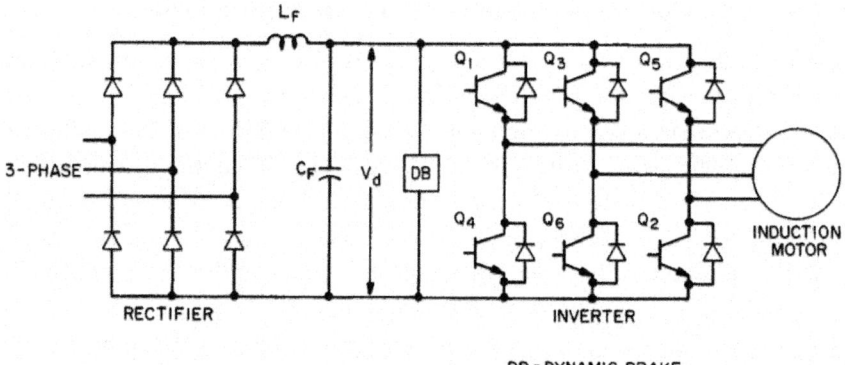

Figure 43: Diode rectifier PWM inverter control of an induction motor.

Often a drive may need only a one- or two-quadrant mode of operation. In such a case, the converter topology can be simple. For example, in one-quadrant drive, only Q_2 chopping and D_3 freewheeling devices are required, and the terminal A is connected to the supply positive. Similarly, a two-quadrant drive will need only one leg of the bridge, where the upper device can be controlled for motoring mode and the lower device can be controlled for regeneration mode.

Converter Control of AC Machines

Although application of dc drives is quite common, disadvantages are that the machines are bulky and expensive, and the commutators and brushes require frequent maintenance. In fact, commutator sparking prevents machine application in an unclean environment, at high speed, and at high elevation. AC machines, particularly the cage-type induction motor, are favorable when compared with all the features of dc machines. Although converter system, control, and signal processing of ac drives is definitely complex, the evolution of ac drive technology in the past two decades has permitted more economical and higher performance ac drives. Consequently, ac drives are finding expanding applications, pushing dc drives towards obsolescence.

Voltage-Fed Inverter Induction Motor Drive

A simple and popular converter system for speed control of an induction motor is shown in Fig. 43. The front-end diode rectifier converts 60 Hz ac to dc, which is then filtered to remove the ripple. The dc voltage is then converted to variable-frequency, variable-voltage output for the machine through a PWM bridge inverter. Among a number of PWM techniques, the sinusoidal

PWM is common, and it is illustrated in Fig. 44 for one phase only. The stator sinusoidal reference phase voltage signal is compared with a high-frequency carrier wave, and the comparator logic output controls switching of the upper and lower transistors in a phase leg. The phase voltage wave shown refers to the fictitious center tap of the filter capacitor. With the PWM technique, the fundamental voltage and frequency can be easily varied. The stator voltage wave contains high-frequency ripple, which is easily filtered by the machine leakage inductance. The voltage-to-frequency ratio is kept constant to provide constant air gap flux in the machine. The machine voltage-frequency relation, and the corresponding torque, stator current, and slip, are shown in Fig. 45. Up to the base or rated frequency wb, the machine can develop constant torque. Then, the field flux weakens as the frequency is increased at constant voltage. The speed of the machine can be controlled in a simple open-loop manner by controlling the frequency and maintaining the proportionality between the voltage and frequency. During acceleration, machine-developed torque should be limited so that the inverter current rating is not exceeded. By controlling the frequency, the operation can be extended in the field weakening region. If the supply frequency is controlled to be lower than the machine speed (equivalent frequency), the motor will act as a generator and the inverter will act as a rectifier, and energy from the motor will be pumped back to the dc link. The dynamic brake shown is nothing but a buck converter with resistive load that dissipates excess power to maintain the dc bus voltage constant. When the motor speed is reduced to zero, the phase sequence of the inverter can be reversed for speed reversal. Therefore, the machine speed can be easily controlled in all four quadrants.

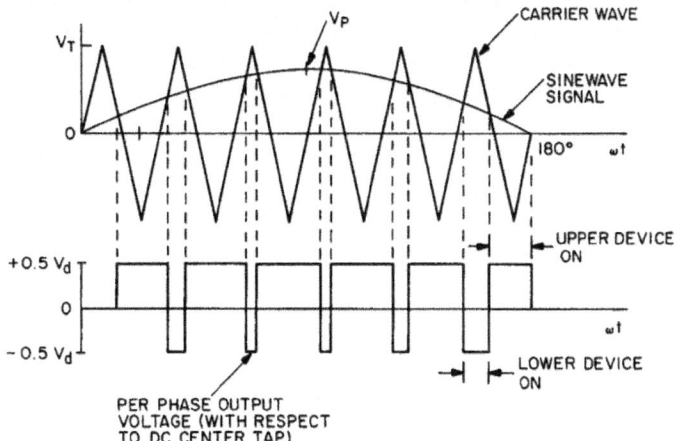

Figure 44: Sinusoidal pulse width modulation principle.

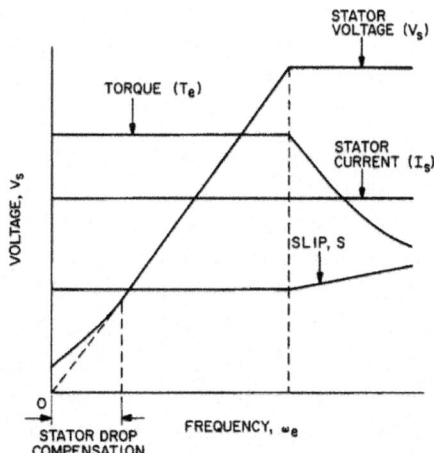

Figure 45: Voltage-frequency relation of an induction motor.

Current-Fed Inverter Induction Motor Drive

The speed of a machine can be controlled by a current-fed inverter as shown in Fig. 46. The front-end thyristor rectifier generates a variable dc current source in the dc link inductor. The dc current is then converted to six-step machine current wave through the inverter. The basic mode of operation of the inverter is the same as that of the rectifier, except that it is force-commutated, that is, the capacitors and series diodes help commutation of the thyristors. One advantage of the drive is that regenerative braking is easy because the rectifier and inverter can reverse their operation modes. Six-step machine current, however, causes large harmonic heating and torque pulsation, which may be quite harmful at low-speed operation. Another disadvantage is that the converter system cannot be controlled in open loop like a voltage-fed inverter.

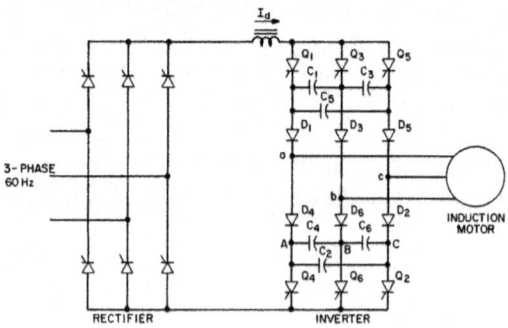

Figure 46: Force-commutated current-fed inverter control of an induction motor.

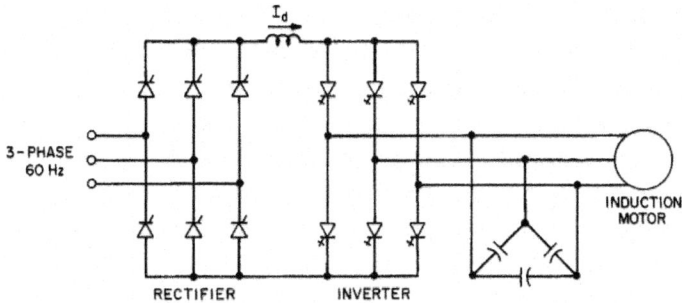

Figure 47: PWM current-fed inverter control of an induction motor.

Current-Fed PWM Inverter Induction Motor Drive

The force-commutated thyristor inverter in Fig. 46 can be replaced by a self-commutating gate turn-off (GTO) thyristor PWM inverter as shown in Fig. 47. The output capacitor bank shown has two functions: (1) it permits PWM switching of the GTO by diverting the load inductive current, and (2) it acts as a low-pass filter causing sinusoidal machine current. The second function improves machine efficiency and attenuates the irritating magnetic noise. Note that the fundamental machine current is controlled by the front-end rectifier, and the fixed PWM pattern is for controlling the harmonics only. The GTO is to be the reverse-blocking type. Such drives are popular in the multimegawatt power range. For lower power, an insulated gate bipolar transistor (IGBT) or transistor can be used with a series diode.

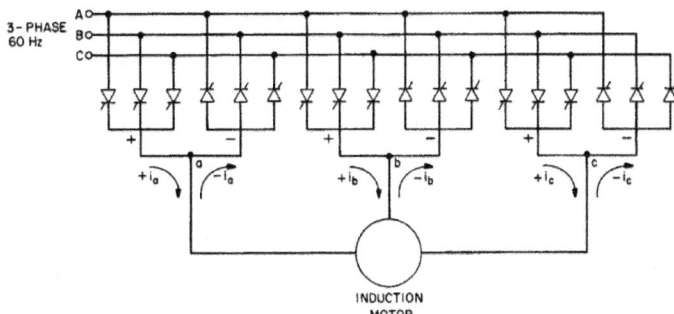

Figure 48: Cycloconverter control of an induction motor.

Cycloconverter Induction Motor Drive

A phase-controlled cycloconverter can be used for speed control of an ac machine (induction or synchronous type). Figure 48 shows a drive using a three-pulse half-wave or 18-thyristor cycloconverter. Each output phase group consists of positive and negative converter components which permit bidirectional current flow. The firing angle of each converter is sinusoidally modulated to generate the variable-frequency, variable-voltage output required for ac machine drive. Speed reversal and regenerative mode operation are easy. The cycloconverter can be operated in blocking or circulating current mode. In blocking mode, the positive or negative converter is enabled, depending on the polarity of the load current. In circulating current mode, the converter components are always enabled to permit circulating current through them. The circulating current reactor between the positive and negative converter prevents short circuits due to ripple voltage. The circulating current mode gives simple control and a higher range of output frequency with lower harmonic distortion.

Slip Power Recovery Drive of Induction Motor

In a cage-type induction motor, the rotor current at slip frequency reacting with the airgap flux develops the torque. The corresponding slip power is dissipated in the rotor resistance. In a wound rotor induction motor, the slip power can be controlled to control the torque and speed of a machine. Figure 49 shows a popular slip power-controlled drive, known as a static Kramer drive. The slip power is rectified to dc with a diode rectifier and is then pumped back to an ac line through a thyristor phase-controlled inverter. The method permits speed control in the subsynchronous speed range. It can be shown that the developed machine torque is proportional to the dc link current I_d and the voltage V_d varies directly with speed deviation from the synchronous speed. The current I_d is controlled by the firing angle of the inverter. Since V_d and V_I voltages balance at steady state, at synchronous speed the voltage V_d is zero and the firing angle is 90 degrees. The firing angle increases as the speed falls, and at 50% synchronous speed the firing angle is near 180 degrees. This is practically the lowest speed in static Kramer drive. The transformer steps down the inverter input voltage to get a 180-degree firing angle at lowest speed. The advantage of this drive is that the converter rating is low compared with the machine rating. Disadvantages are that the line power factor is low and the machine is expensive. For limited speed range applications, this drive has been popular.

Wound Field Synchronous Motor Drive

The speed of a wound field synchronous machine can be controlled by a current-fed converter scheme as shown in Fig. 46, except that the forced-commutation elements can be removed. The machine is operated at leading power factor by overexcitation so that the inverter can be load commutated. Because of the simplicity of converter topology and control, such a drive is popular in the multimegawatt range.

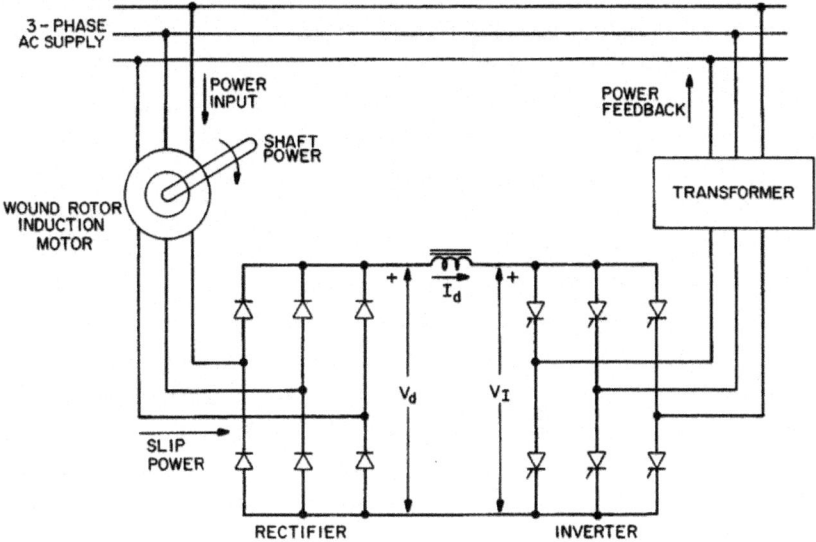

Figure 49: Slip power recovery control of a wound rotor induction motor.

Permanent Magnet Synchronous Motor Drive

Permanent magnet (PM) machine drives are quite popular in the low-power range. A PM machine can have sinusoidal or concentrated winding, giving the corresponding sinusoidal or trapezoidal induced stator voltage wave. Figure 50 shows the speed control system using a trapezoidal machine, and Fig. 51 explains the wave forms. The power MOSFET inverter supplies variable-frequency, variable-magnitude six-step current wave to the stator. The inverter is self-controlled, that is, the firing pulses are generated by the machine position sensor through a decoder. It can be shown that such a drive has the features of dc drive and is normally defined as brushless dc drive. The speed control loop generates the dc current command, which is then controlled by the hysteresis-band method to construct the six-step phase current waves in correct phase relation with the induced voltage waves as shown in Fig. 51. The drive can easily operate in four-quadrant mode.

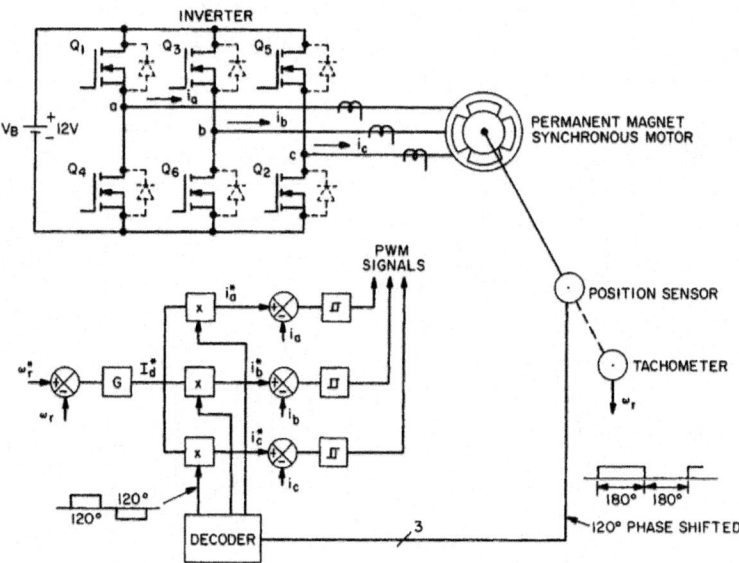

Figure 50: Permanent magnet synchronous motor control with PWM inverter.

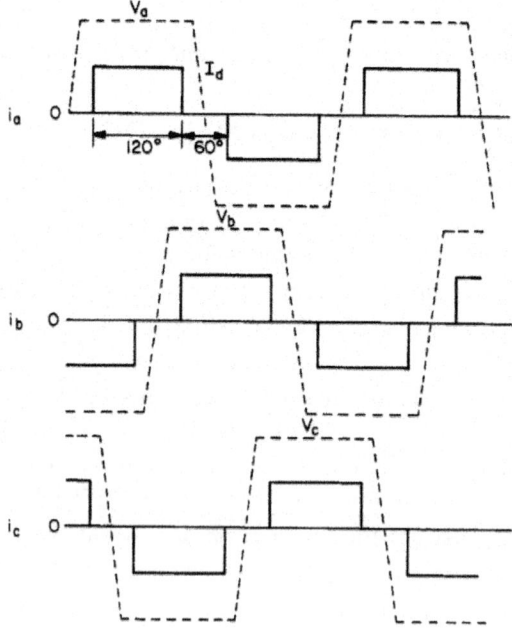

Figure 51: Phase voltage and current waves in brushless dc drive.

REFERENCES

1. B.K. Bose, Modern Power Electronics: Evaluation, Technology, and Applications, New York: IEEE Press, 1992. Harris Semiconductor, User's Guide of MOS Controlled Thyristor.

2. A.Q. Huang, Recent Developments of Power Semiconductor Devices, VPEC Seminar Proceedings, pp. 1–9, September 1995.

3. N. Mohan and T. Undeland, Power Electronics: Converters, Applications, and Design, New York: John Wiley & Sons, 1995.

4. J. Wojslawowicz, "Ruggedized transistors emerging as power MOSFET standard-bearers," Power Technics Magazine, pp. 29–32, January 1988.

5. B.K. Bose, Modern Power Electronics, New York: IEEE Press, 1992.

6. Motorola, Linear/Switchmode Voltage Regulator Handbook, 1989.

7. K.S. Rajashekara, H. Le-Huy, et al., "Resonant DC Link Inverter-Fed AC Machines Control," IEEE Power Electronics Specialists Conference, 1987, pp. 491–496.

8. P.C. Sen, Thyristor DC Drives, New York: John Wiley, 1981.

9. G. Venkataramanan and D. Divan, "Pulse Width Modulation with Resonant DC Link Converters," IEEE IAS Annual Meeting, 1990, pp. 984–990.

10. A.K.S. Bhat, "A unified approach for the steady-state analysis of resonant converters," IEEE Trans. Industrial Electronics, vol. 38, no. 4, pp. 251–259, Aug. 1991.

11. A.K.S. Bhat, "Fixed frequency PWM series-parallel resonant converter," IEEE Trans. Industry Applications, vol. 28, no. 5, pp. 1002–1009, 1992.

12. E.R. Hnatek, Design of Solid-State Power Supplies, 2nd ed., New York: Van Nostrand Reinhold, 1981.

13. K.H. Liu and F.C. Lee, "Zero-Voltage Switching Technique In DC/DC Converters," IEEE Power Electronics Specialists Conference Record, 1986, pp. 58–70.

14. K.H. Liu, R. Oruganti, and F.C. Lee, "Resonant Switches—Topologies and Characteristics," IEEE Power Electronics Specialists Conference Record, 1985, pp. 106–116.

15. Motorola, Linear/Switchmode Voltage Regulator Handbook, 1989.

16. Philips Semiconductors, Power Semiconductor Applications, 1991.

17. M.H. Rashid, Power Electronics: Circuits, Devices, and Applications, Englewood Cliffs, N.J.: Prentice-Hall, 1988.

18. R. Severns and G. Bloom, Modern Switching DC-to-DC Converters,

New York: Van Nostrand Reinhold, 1988.

19. R.L. Steigerwald, "A comparison of half-bridge resonant converter topologies," IEEE Trans. Power Electron., vol. PE-3, no. 2, pp. 174–182, April 1988.

20. K.K. Sum, Recent Developments in Resonant Power Conversion, Calif.: Intertech Communications, 1988.

21. Unitrode Switching Regulated Power Supply Design Seminar Manual, Lexington, Mass.: Unitrode Corporation, 1984.

22. B.K. Bose, Power Electronics and AC Drives, Englewood Cliffs, N.J.: Prentice-Hall, 1986.

23. B.K. Bose, "Adjustable speed AC drives—A technology status review," Proc. IEEE, vol. 70, pp. 116–135, Feb. 1982.

24. B.K. Bose, Modern Power Electronics, New York: IEEE Press, 1992.

25. J.M.D. Murphy and F.G. Turnbull, Power Electronic Control of AC Motors, New York: Pergamon Press, 1988.

26. P.C. Sen, Thyristor DC Drives, New York: John Wiley, 1981.

Chapter 6

THREE-PHASE CONTROLLED RECTIFIERS

INTRODUCTION

Three-phase controlled rectifiers have a wide range of applications, from small rectifiers to large high voltage direct current (HVDC) transmission systems. They are used for electrochemical processes, many kinds of motor drives, traction equipment, controlled power supplies and many other applications. From the point of view of the commutation process, they can be classified into two important categories: line-commutated controlled rectifiers (thyristor rectifiers) and force-commutated pulse width modulated (PWM) rectifiers.

LINE-COMMUTATED CONTROLLED RECTIFIERS

Three-phase Half-wave Rectifier

Figure 1 shows the three-phase half-wave rectifier topology. To control the load voltage, the half-wave rectifier uses three common-cathode thyristor arrangement. In this figure, the power supply and the transformer are assumed ideal. The thyristor will conduct (ON state), when the anode-to-cathode voltage v_{AK} is positive and a firing current pulse i_G is applied to the gate terminal. Delaying the firing pulse by an angle α controls the load voltage. As shown in Fig. 2, the firing angle α is measured from the crossing point between the phase supply voltages. At that point, the anode-to-cathode thyristor voltage v_{AK} begins to be positive. Figure 3 shows that the possible range for gating delay is between $\alpha = 0°$ and $\alpha = 180°$, but because of commutation problems in actual situations, the maximum firing angle is limited to around 160°. As shown in Fig. 4, when the load is resistive, current i_d has the same waveform of the load voltage. As the load becomes more and more inductive, the current flattens and finally becomes constant. The thyristor goes to the non-conducting condition (OFF state) when the following thyristor is switched ON, or the current, tries to reach a negative value. With the help of Fig. 2, the load average voltage can be evaluated, and is given by:

$$V_D = \frac{V_{MAX}}{\frac{2}{3}\pi} \int\limits_{-\pi/3+\alpha}^{\pi/3+\alpha} \cos\omega t \cdot d(\omega t)$$

$$= V_{MAX} \frac{\sin\frac{\pi}{3}}{\frac{\pi}{3}} \cdot \cos\alpha \approx 1.17 \cdot V_{f-N}^{rms} \cdot \cos\alpha$$

(1)

where V_{MAX} is the secondary phase-to-neutral peak voltage, V_{f-N}^{rms} its root mean square (rms) value and ω is the angular frequency of the main power supply. It can be seen from Eq. (1) that the load average voltage V_D is modified by changing firing angle α. When $\alpha < 90°$, V_D is positive and when $\alpha > 90°$, the average dc voltage becomes negative.

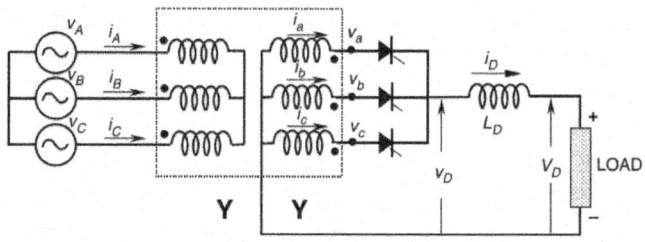

Figure 1: Three-phase half-wave rectifier.

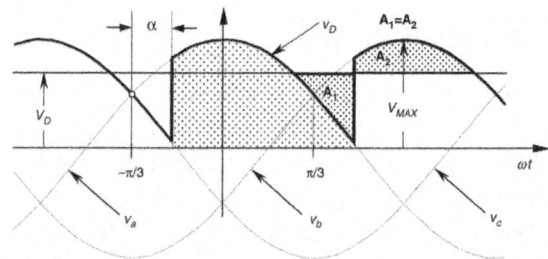

Figure 2: Instantaneous dc voltage v_D, average dc voltage V_D, and firing angle α.

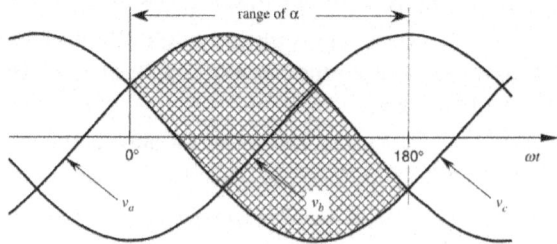

Figure 3: Possible range for gating delay in angle α.

In such a case, the rectifier begins to work as an inverter and the load needs to be able to generate power reversal by reversing its dc voltage.

The ac currents of the half-wave rectifier are shown in Fig. 5. This drawing assumes that the dc current is constant (very large L_D). Disregarding commutation overlap, each valve conducts during 120° per period. The secondary currents (and thyristor currents) present a dc component that is undesirable, and makes this rectifier not useful for high power applications.

The primary currents show the same waveform, but with the dc component removed. This very distorted waveform requires an input filter to reduce harmonics contamination.

The current waveforms shown in Fig. 5 are useful for designing the power transformer. Starting from:

$$VA_{prim} = 3 \cdot V_{(prim)f-N}^{rms} \cdot I_{prim}^{rms}$$

$$VA_{sec} = 3 \cdot V_{(sec)f-N}^{rms} \cdot I_{sec}^{rms}$$

$$P_D = V_D \cdot I_D$$

$$(2)$$

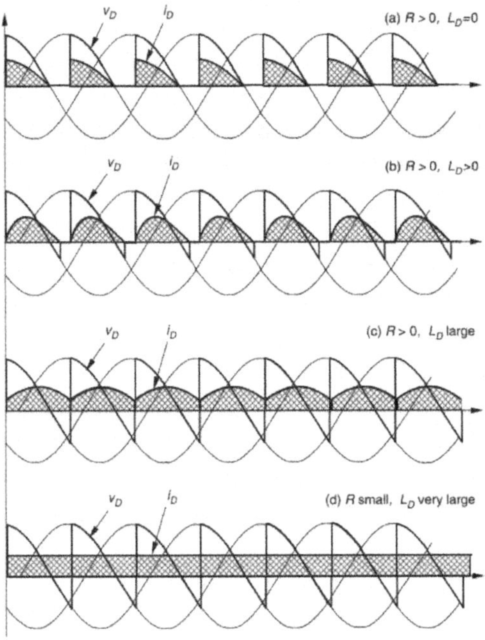

Figure 4: DC current waveforms.

where VA_{prim} and VA_{sec} are the ratings of the transformer for the primary and secondary side respectively. Here P_D is the power transferred to the dc side. The maximum power transfer is with $\alpha = 0°$ (or $\alpha = 180°$). Then, to establish a relation between ac and dc voltages, Eq. (1) for $\alpha = 0°$ is required:

$$V_D = 1.17 \cdot V^{rms}_{(sec)f-N} \tag{3}$$

$$V_D = 1.17 \cdot a \cdot V^{rms}_{(prim)f-N} \tag{4}$$

where a is the secondary to primary turn relation of the transformer. On the other hand, a relation between the currents is also possible to obtain. With the help of Fig. 5:

$$I^{rms}_{sec} = \frac{I_D}{\sqrt{3}} \tag{5}$$

$$I^{rms}_{prim} = a \cdot \frac{I_D\sqrt{2}}{3} \tag{6}$$

Combining Eqs. (2) to (6), it yields:

$$VA_{prim} = 1.21 \cdot P_D$$

$$VA_{sec} = 1.48 \cdot P_D \tag{7}$$

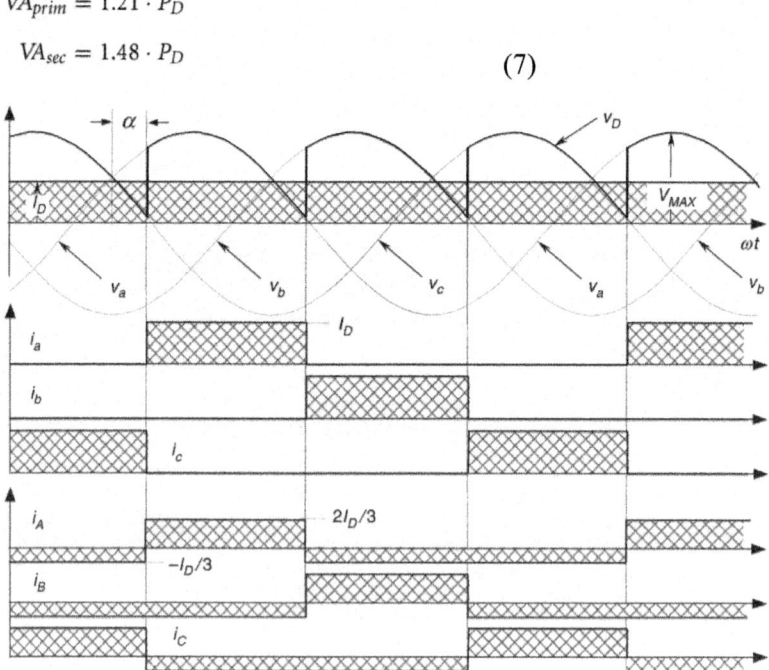

Figure 5: AC current waveforms for the half-wave rectifier.

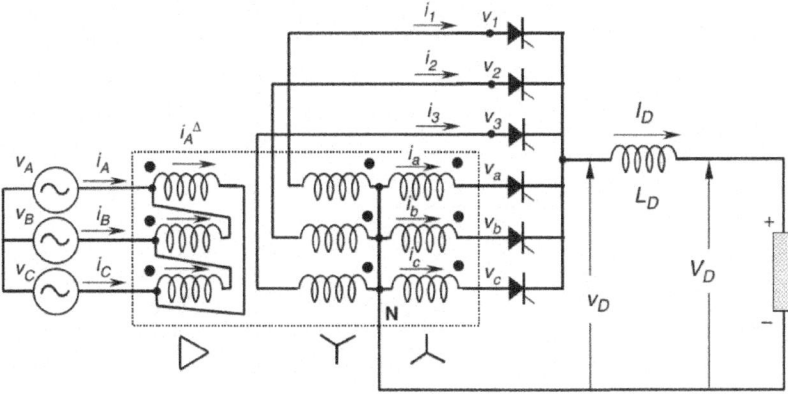

Figure 6: Six-pulse rectifier.

Equation (7) shows that the power transformer has to be oversized 21% at the primary side, and 48% at the secondary side. Then, a special transformer has to be built for this recti- fier. In terms of average VA, the transformer needs to be 35% larger that the rating of the dc load. The larger rating of the secondary with respect to primary is because the secondary carries a dc component inside the windings. Furthermore, the transformer is oversized because the circulation of current harmonics does not generate active power. Core saturation, due to the dc components inside the secondary windings, also needs to be taken in account for iron oversizing.

Six-pulse or Double Star Rectifier

The thyristor side windings of the transformer shown in Fig. 6 form a six-phase system, resulting in a six-pulse starpoint (midpoint connection). Disregarding commutation overlap, each valve conducts only during 60° per period. The direct voltage is higher than that from the half-wave rectifier and its average value is given by:

$$V_D = \frac{V_{MAX}}{\frac{\pi}{3}} \int_{-\pi/6+\alpha}^{\pi/6+\alpha} \cos\omega t \cdot d\,(\omega t)$$

$$= V_{MAX}\frac{\sin\frac{\pi}{6}}{\frac{\pi}{6}} \cdot \cos\alpha \approx 1.35 \cdot V_{f-N}^{rms} \cdot \cos\alpha \tag{8}$$

The dc voltage ripple is also smaller than the one generated by the half-wave rectifier, due to the absence of the third harmonic with its inherently high amplitude. The smoothing reactor L_D is also considerably smaller than the one needed for a three-pulse (half-wave) rectifier.

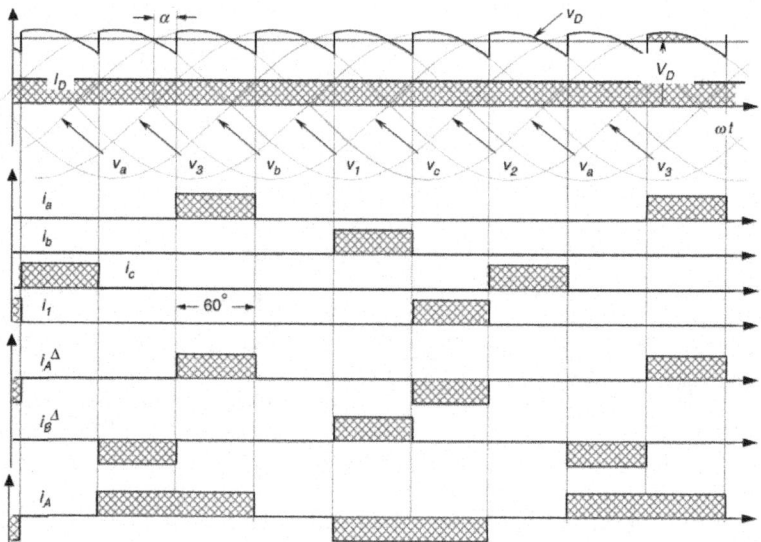

Figure 7: AC current waveforms for the six-pulse rectifier.

The ac currents of the six-pulse rectifier are shown in Fig. 7. The currents in the secondary windings present a dc component, but the magnetic flux is compensated by the double star. As can be observed, only one valve is fired at a time and then this connection in no way corresponds to a parallel connection. The currents inside the delta show a symmetrical waveform with 60° conduction. Finally, due to the particular transformer connection shown in Fig. 6, the source currents also show a symmetrical waveform, but with 120° conduction.

Evaluation of the rating of the transformer is done in similar fashion to the way the half-wave rectifier is evaluated:

$$VA_{prim} = 1.28 \cdot P_D$$

$$VA_{sec} = 1.81 \cdot P_D \qquad (9)$$

Thus the transformer must be oversized 28% at the primary side and 81% at the secondary side. In terms of size it has an average apparent power of 1.55 times the power P_D(55% oversized). Because of the short conducting period of the valves, the transformer is not particularly well utilized.

Double Star Rectifier with Interphase Connection

This topology works as two half-wave rectifiers in parallel, and is very useful when high dc current is required. An optimal way to reach both good balance and elimination of harmonics is through the connection shown in Fig. 8. The two rectifiers are shifted by 180°, and their secondary neutrals are connected through a middle-point autotransformer called "interphase transformer." The interphase transformer is connected between the two secondary neutrals and the middle point at the load return. In this way, both groups operate in parallel. Half the direct current flows in each half of the interphase transformer, and then its iron core does not become saturated. The potential of each neutral can oscillate independently, generating an almost triangular voltage waveform (v_T) in the interphase transformer, as shown in Fig. 9. As this converter work like two half-wave rectifiers connected in parallel, the load average voltage is the same as in Eq. (1):

$$V_D \approx 1.17 \cdot V_{f-N}^{rms} \cdot \cos \alpha$$

$$(10)$$

where V_{f-N}^{rms} is the phase-to-neutral rms voltage at the valve side of the transformer (secondary).

The Fig. 9 also shows the two half-wave rectifier voltages, related to their respective neutrals. Voltage v_{D1} represents the potential between the common cathode connection and the neutral N1. The voltage v_{D2} is between the common cathode connection and N2. It can be seen that the two instantaneous voltages are shifted, which gives as a result, a voltage v_D that is smoother than v_{D1} and v_{D2}.

Figure 10 shows how v_D, v_{D1}, v_{D2}, and v_T change when the firing angle changes from $\alpha = 0°$ to $180°$.

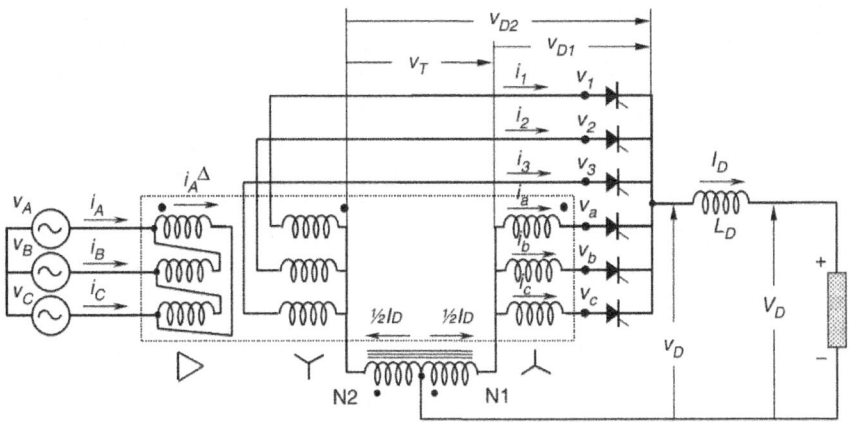

Figure 8: Double star rectifier with interphase transformer.

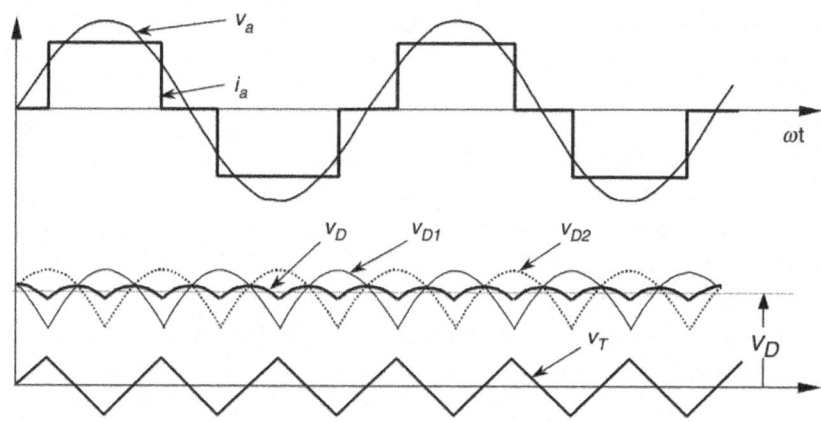

Figure 9: Operation of the interphase connection for $\alpha = 0°$.

The transformer rating in this case is:

$$VA_{prim} = 1.05 \cdot P_D$$

$$VA_{sec} = 1.48 \cdot P_D \tag{11}$$

And the average rating power will be 1.26 P_D, which is better than the previous rectifiers (1.35 for the half-wave rectifier and 1.55 for the six-pulse rectifier). Thus the transformer is well utilized. Figure 11 shows ac current waveforms for a rectifier with interphase transformer.

Three-phase Full-wave Rectifier or Graetz Bridge

Parallel connection via interphase transformers permits the implementation of rectifiers for high current applications. Series connection for high voltage is also possible, as shown in the full-wave rectifier of Fig. 12. With this arrangement, it can be seen that the three common cathode valves generate a positive voltage with respect to the neutral, and the threecommon anode valves produce a negative voltage. The result is a dc voltage, twice the value of the half-wave rectifier. Each half of the bridge is a three-pulse converter group. This bridge connection is a two-way connection and alternating currents flow in the valve-side transformer windings during both half periods, avoiding dc components into the windings, and saturation in the transformer magnetic core. These characteristics make the so-called Graetz bridge the most widely used linecommutated thyristor rectifier. The configuration does not need any special transformer and works as a six-pulse rectifier. The series characteristic of this rectifier produces a dc voltage twice the value of the half-wave rectifier.

The load average voltage is given by:

$$V_D = \frac{2 \cdot V_{MAX}}{\frac{2}{3}\pi} \int\limits_{-\pi/3+\alpha}^{\pi/3+\alpha} \cos\omega t \cdot d(\omega t)$$

$$= 2 \cdot V_{MAX} \frac{\sin\frac{\pi}{3}}{\frac{\pi}{3}} \cdot \cos\alpha \approx 2.34 \cdot V_{f-N}^{rms} \cdot \cos\alpha$$

$$(12)$$

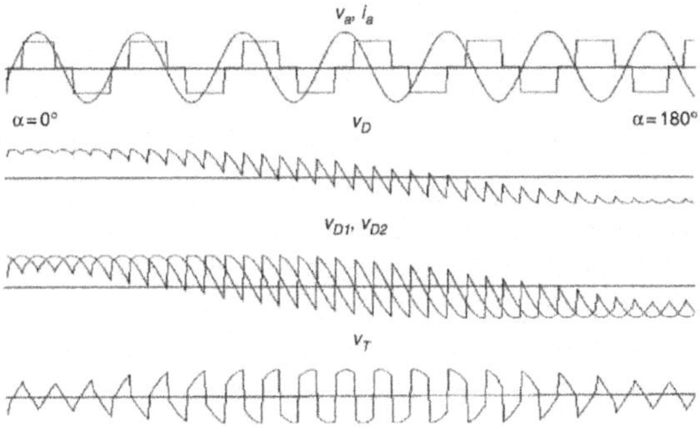

Figure 10: Firing angle variation from $\alpha = 0°$ to $180°$.

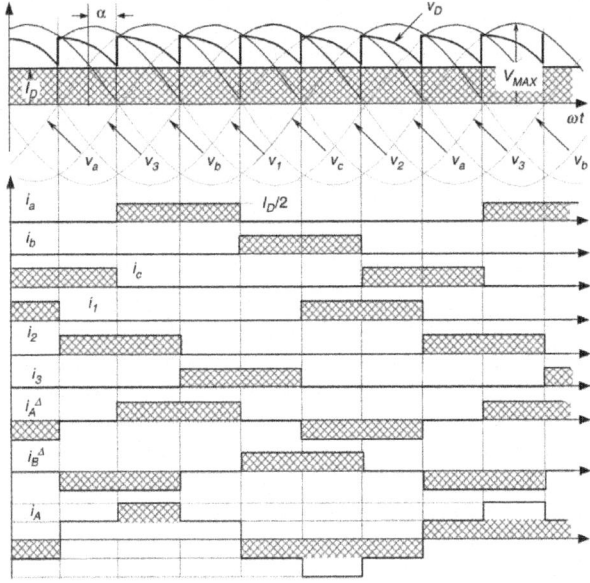

Figure 11: AC current waveforms for the rectifier with interphase transformer.

Or

$$V_D = \frac{3 \cdot \sqrt{2} \cdot V_{f-f}^{sec}}{\pi} \cos \alpha \approx 1.35 \cdot V_{f-f}^{sec} \cdot \cos \alpha$$

(13)

where V_{MAX} is the peak phase-to-neutral voltage at the secondary transformer terminals, V_{f-N}^{rms} its rms value, and V_{f-f}^{sec} the rms phase-to-phase secondary voltage, at the valve terminals of the rectifier.

Figure 13 shows the voltages of each half-wave bridge of this topology, v_D^{pos} and v_D^{neg}, the total instantaneous dc voltage v_D, and the anode-to-cathode voltage v_{AK} in one of the bridge thyristors. The maximum value of v_{AK} is $\sqrt{3} \cdot V_{MAX}$, which is the same as that of the half-wave converter and the interphase transformer rectifier.

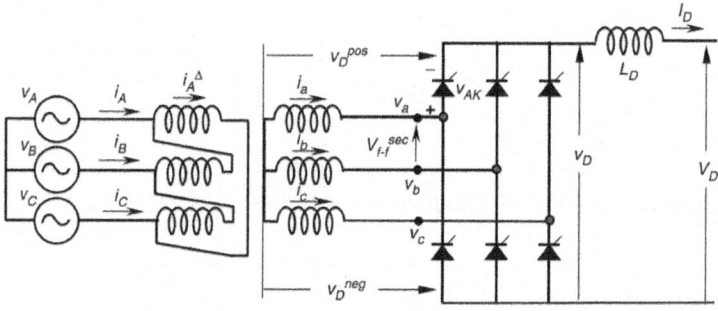

Figure 12: Three-phase full-wave rectifier or Graetz bridge.

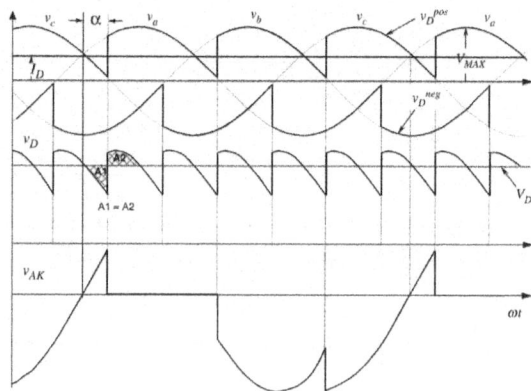

Figure 13: Voltage waveforms for the Graetz bridge.

The double star recti- fier presents a maximum anode-to-cathode voltage of two times V_{MAX}. Figure 14 shows the currents of the rectifier, which assumes that L_D is large enough to keep the dc current smooth. The example is for the same Y transformer connection shown in the topology of Fig. 12. It can be noted that the secondary currents do not carry any dc component, thereby avoiding overdesign of the windings and transformer saturation. These two figures have been drawn for a firing angle, α of approximately 30°. The perfect symmetry of the currents in all windings and lines is one of the reasons why this rectifier is the most popular of its type. The transformer rating in this case is

$$VA_{prim} = 1.05 \cdot P_D$$

$$VA_{sec} = 1.05 \cdot P_D \qquad (14)$$

As can be noted, the transformer needs to be oversized only 5%, and both primary and secondary windings have the same rating. Again, this value can be compared with the previous rectifier transformers: $1.35P_D$ for the half-wave rectifier, $1.55P_D$ for the six-pulse rectifier, and $1.26P_D$ for the interphase transformer rectifier. The Graetz bridge makes excellent use of the power transformer.

Half Controlled Bridge Converter

The fully controlled three-phase bridge converter shown in Fig. 12 has six thyristors. As already explained here, this circuit operates as a rectifier when each thyristor has a firing angle $\alpha < 90°$ and functions as an inverter for $\alpha > 90°$. If inverter operation is not required, the circuit may be simplified by replacing three controlled rectifiers with power diodes, as in Fig. 15a. This simplification is economically attractive because diodes are considerably less expensive than thyristors and they do not require firing angle control electronics.

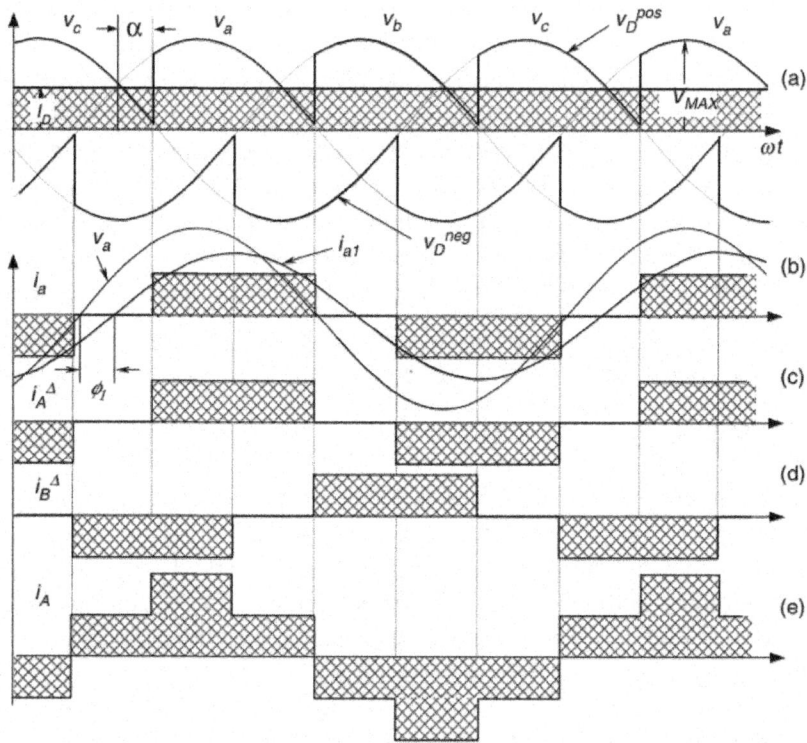

Figure 14: Current waveforms for the Graetz bridge.

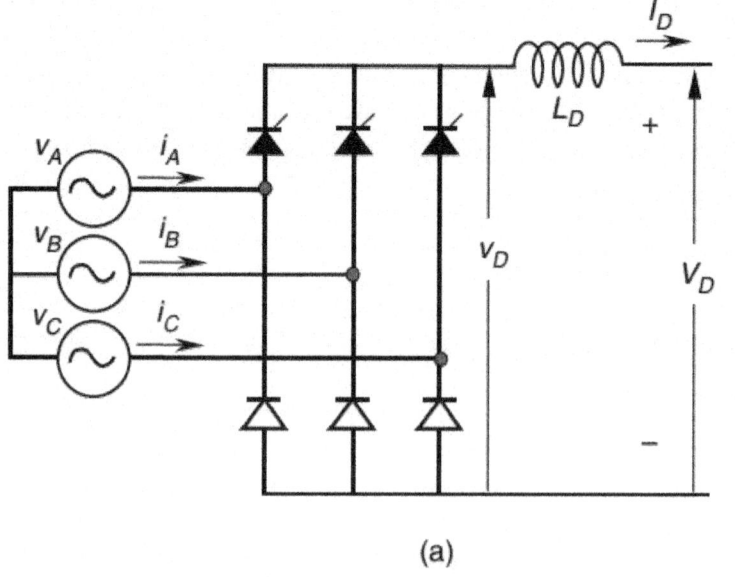

(a)

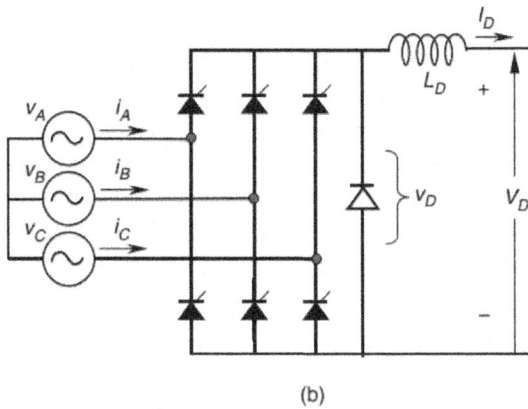

(b)

Figure 15: One-quadrant bridge converter circuits: (a) half-controlled bridge and (b) free-wheeling diode bridge.

The half controlled bridge, or "semiconverter," is analyzed by considering it as a phase-controlled half-wave circuit in series with an uncontrolled half-wave rectifier. The average dc voltage is given by the following equation

$$V_D = \frac{3 \cdot \sqrt{2} \cdot V_{f-f}^{sec}}{2\pi}(1 + \cos\alpha)$$

(15)

Then, the average voltage V_D never reaches negative values. The output voltage waveforms of half-controlled bridge are similar to those of a fully controlled bridge with a freewheeling diode. The advantage of the free-wheeling diode connection, shown in Fig. 15b is that there is always a path for the dc current, independent of the status of the ac line and of the converter. This can be important if the load is inductive–resistive with a large time constant, and there is an interruption in one or more of the line phases. In such a case, the load current could commutate to the free-wheeling diode.

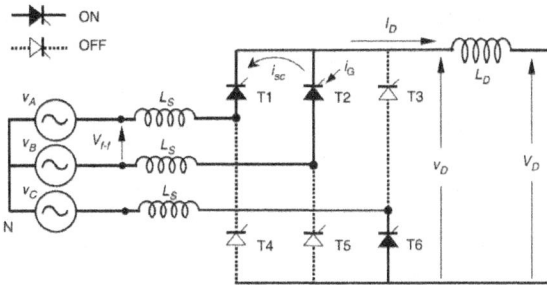

Figure 16: Commutation process.

Commutation

The description of the converters in the previous sections was based upon assumption that the commutation was instantaneous. In practice this is not possible, because the transfer of current between two consecutive valves in a commutation group takes a finite time. This time, called overlap time, depends on the phase-to-phase voltage between the valves participating in the commutation process, and the line inductance L_S between the converter and power supply. During the overlap time, two valves conduct, and the phase-to-phase voltage drops entirely on the inductances L_S. Assuming the dc current I_D to be smooth and with the help of Fig. 16, the following relation is deduced

$$2L_S \cdot \frac{di_{sc}}{dt} = \sqrt{2} \cdot V_{f-f} \sin \omega t = v_A - v_B \tag{16}$$

where i_{sc} is the current in the valve being fired during the commutation process (thyristor T2 in Fig. 16). This current can be evaluated, and it yields

$$i_{sc} = -\frac{\sqrt{2}}{2L_S} \cdot V_{f-f} \frac{\cos \omega t}{\omega} + C \tag{17}$$

The constant "C" is evaluated through initial conditions at the instant when T2 is ignited. In terms of angle, when $\omega t = \alpha$

when $\omega t = \alpha$, $i_{sc} = 0$

$$\therefore C = \frac{V_{f-f}^{sec}}{\sqrt{2} \cdot \omega L_S} \cos \alpha \tag{18}$$

Replacing Eq. (18) in (17):

$$i_{sc} = \frac{V_{f-f}}{\sqrt{2} \cdot \omega L_S} \cdot (\cos \alpha - \cos \omega t) \tag{19}$$

Before commutation, the current I_D was carried by thyristor T1 (see Fig. 16). During the commutation time, the load current I_D remains constant, i_{sc} returns through T1, and T1 is automatically switched off when the current i_{sc} reaches the value of I_D. This happens because thyristors cannot conduct in reverse direction. At this moment, the overlap time lasts and the current I_D is then conducted by T2. In terms of angle, when $\omega t = \alpha + \mu$, $i_{sc} = I_D$, where μ is defined as the "overlap angle." Replacing this final condition in Eq. (19) yields

$$I_D = \frac{V_{f-f}^{sec}}{\sqrt{2} \cdot \omega L_S} \cdot [\cos \alpha - \cos (\alpha + \mu)] \tag{20}$$

To avoid confusion in a real analysis, it has to be remembered that V_{f-f} corresponds to the secondary voltage in case of transformer utilization. For this reason, the abbreviation "sec" has been added to the phase-to-phase voltage in Eq. (20).

During commutation, two valves conduct at a time, which means that there is an instantaneous short circuit between the two voltages participating in the process. As the inductances of each phase are the same, the current i_{sc} produces the same voltage drop in each L_S , but with opposite sign because this current flows in reverse direction and with opposite slope in each inductance. The phase with the higher instantaneous voltage suffers a voltage drop $-\Delta v$ and the phase with the lower voltage suffers a voltage increase $+\Delta v$. This situation affects the dc voltage VC, reducing its value an amount ΔV_{med} . Figure 17 shows the meanings of Δv, ΔV_{med} , μ, and i_{sc}.

The area ΔV_{med} showed in Fig. 17, represents the loss of voltage that affects the average voltage V_C, and can be evaluated through the integration of Δv during the overlap angle μ. The voltage drop Δv can be expressed as

$$\Delta v = \left(\frac{v_A - v_B}{2} \right) = \frac{\sqrt{2} \cdot V_{f-f}^{sec} \sin \omega t}{2}$$

(21)

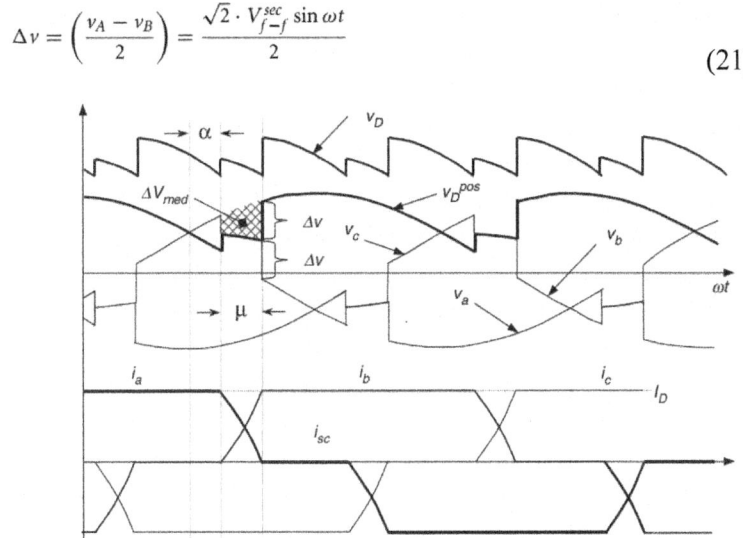

Figure 17: Effect of the overlap angle on the voltages and currents.

Integrating Eq. (21) into the corresponding period (60°) and interval (μ) and starting at the instant when the commutation begins (α)

$$\Delta V_{med} = \frac{3}{\pi} \cdot \frac{1}{2} \int_{\alpha}^{\alpha+\mu} \sqrt{2} \cdot V_{f-f}^{sec} \sin \omega t \cdot d\omega t$$

(22)

$$\Delta V_{med} = \frac{3 \cdot V_{f-f}^{sec}}{\pi \cdot \sqrt{2}} \left[\cos \alpha - \cos (\alpha + \mu)\right]$$

(23)

Subtracting ΔV_{med} in Eq. (13)

$$V_D = \frac{3 \cdot \sqrt{2} \cdot V_{f-f}^{sec}}{\pi} \cos \alpha - \Delta V_{med}$$

(24)

$$V_D = \frac{3 \cdot \sqrt{2} \cdot V_{f-f}^{sec}}{2\pi} \left[\cos \alpha + \cos (\alpha + \mu)\right]$$

(25)

or

$$V_D = \frac{3 \cdot \sqrt{2} \cdot V_{f-f}^{sec}}{\pi} \left[\cos \left(\alpha + \frac{\mu}{2}\right) \cos \frac{\mu}{2}\right]$$

(26)

Equations (20) and (25) can be written as a function of the primary winding of the transformer, if there is any transformer.

$$I_D = \frac{a \cdot V_{f-f}^{prim}}{\sqrt{2} \cdot \omega L_S} \cdot \left[\cos \alpha - \cos (\alpha + \mu)\right]$$

(27)

$$V_D = \frac{3 \cdot \sqrt{2} \cdot a \cdot V_{f-f}^{prim}}{2\pi} \left[\cos \alpha + \cos (\alpha + \mu)\right]$$

(28)

where $a = V_{f-f}^{sec} / V_{f-f}^{prim}$. With Eqs. (27) and (28) one gets

$$V_D = \frac{3 \cdot \sqrt{2}}{\pi} \cdot a \cdot V_{f-f}^{prim} \cos \alpha - \frac{3I_D \omega L_S}{\pi}$$

(29)

Equation (29) allows a very simple equivalent circuit of the converter to be made, as shown in Fig. 18. It is important to note that the equivalent resistance of this circuit is not real, because it does not dissipate power.

From the equivalent circuit, regulation curves for the rectifier under different firing angles are shown in Fig. 19. It should be noted that these curves correspond only to an ideal situation, but helps in understanding the effect of voltage drop Δv on dc voltage. The commutation process and the overlap angle also affects the voltage v_a and anode-to-cathode thyristor voltage, as shown in Fig. 20.

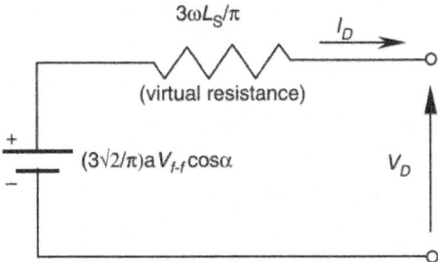

Figure 18: Equivalent circuit for the converter.

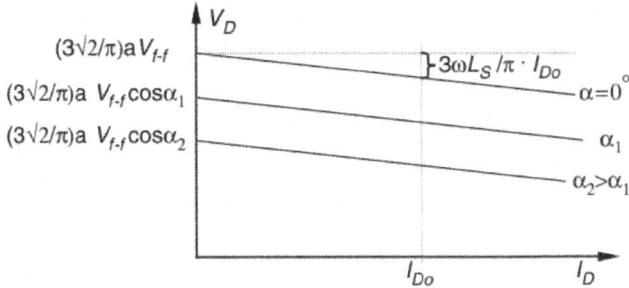

Figure 19: DC voltage regulation curves for rectifier operation.

Power Factor

The displacement factor of the fundamental current, obtained from Fig. 14 is

$$\cos \phi_1 = \cos \alpha \qquad (30)$$

In the case of non-sinusoidal current, the active power delivered per phase by the sinusoidal supply is

$$P = \frac{1}{T} \int_0^T v_a(t) i_a(t) dt = V_a^{rms} I_{a1}^{rms} \cos \phi_1 \qquad (31)$$

where V_a^{rms} is the rms value of the voltage v_a and I_{a1}^{rms} the rms value of i_{a1} (fundamental component of ia). Analog relations can be obtained for v_b and v_c.

The apparent power per phase is given by

$$S = V_a^{rms} I_a^{rms} \qquad (32)$$

The power factor is defined by

$$PF = \frac{P}{S}$$

(33)

By substituting Eqs. (30), (31), and (32) into Eq. (33), the power factor can be expressed as follows

$$PF = \frac{I_{a1}^{rms}}{I_a^{rms}} \cos \alpha$$

(34)

This equation shows clearly that due to the non-sinusoidal waveform of the currents, the power factor of the rectifier is negatively affected by both the firing angle α and the distortion of the input current. In effect, an increase in the distortion of the current produces an increase in the value of I_a^{rms} in Eq. (34), which deteriorates the power factor.

Harmonic Distortion

The currents of the line-commutated rectifiers are far from being sinusoidal. For example, the currents generated from the Graetz rectifier (see Fig. 14b) have the following harmonic content

$$i_A = \frac{2\sqrt{3}}{\pi} I_D (\cos \omega t - \frac{1}{5} \cos 5\omega t + \frac{1}{7} \cos 7\omega t$$

$$- \frac{1}{11} \cos 11\omega t + \ldots)$$

(35)

Some of the characteristics of the currents, obtained from Eq. (35) include: (i) the absence of triple harmonics; (ii) the presence of harmonics of order 6k ± 1 for integer values of k; (iii) those harmonics of orders 6k +1 are of positive sequence; (iv) those of orders 6k − 1 are of negative sequence; (v) the rms magnitude of the fundamental frequency is

$$I_1 = \frac{\sqrt{6}}{\pi} I_D$$

(36)

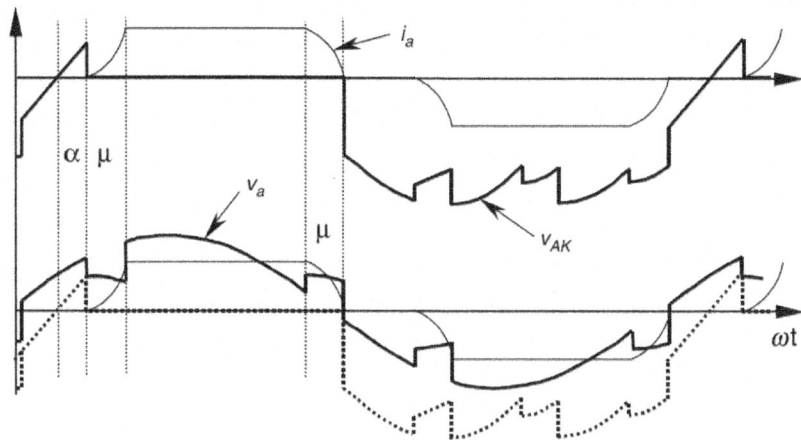

Figure 20: Effect of the overlap angle on v_a and on thyristor voltage v_{AK}.

and (vi) the rms magnitude of the nth harmonic is

$$I_n = \frac{I_1}{n}$$

(37)

If either, the primary or the secondary three-phase windings of the rectifier transformer are connected in delta, the ac side current waveforms consist of the instantaneous differences between two rectangular secondary currents 120° apart as shown in Fig. 14e. The resulting Fourier series for the current in phase "a" on the primary side is

$$i_A = \frac{2\sqrt{3}}{\pi} I_D (\cos \omega t + \frac{1}{5} \cos 5\omega t - \frac{1}{7} \cos 7\omega t$$

$$- \frac{1}{11} \cos 11\omega t + ...)$$

(38)

This series differs from that of a star connected transformer only by the sequence of rotation of harmonic orders $6k \pm 1$ for odd values of k, i.e. the 5th, 7th, 17th, 19th, etc.

Special Configurations for Harmonic Reduction

A common solution for harmonic reduction is through the connection of passive filters, which are tuned to trap a particular harmonic frequency. A typical configuration is shown in Fig. 21.

However, harmonics also can be eliminated using special configurations of converters. For example, 12-pulse configuration consists of two sets of converters connected as shownin Fig. 22. The resultant ac current is given by the sum of the two Fourier series of the star connection (Eq. (35)) and delta connection transformers (Eq. (38)).

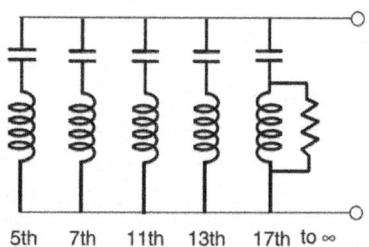

5th   7th   11th   13th   17th to ∞

Figure 21: Typical passive filter for one phase.

$$i_A = 2\left(\frac{2\sqrt{3}}{\pi}\right) I_D (\cos\omega t - \frac{1}{11}\cos 11\omega t + \frac{1}{13}\cos 13\omega t$$

$$-\frac{1}{23}\cos 23\omega t + \ldots)$$

$$(39)$$

The series only contains harmonics of order $12k \pm 1$. The harmonic currents of orders $6k \pm 1$ (with k odd), i.e. 5th, 7th, 17th, 19th, etc. circulate between the two converter transformers but do not penetrate the ac network.

The resulting line current for the 12-pulse rectifier, shown in Fig. 23, is closer to a sinusoidal waveform than previous line currents. The instantaneous dc voltage is also smoother with this connection.

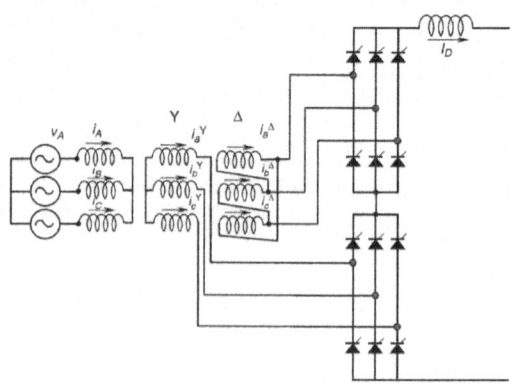

Figure 22: 12-pulse rectifier configuration.

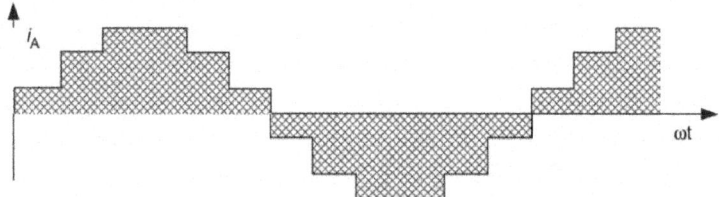

Figure 23: Line current for the 12-pulse rectifier.

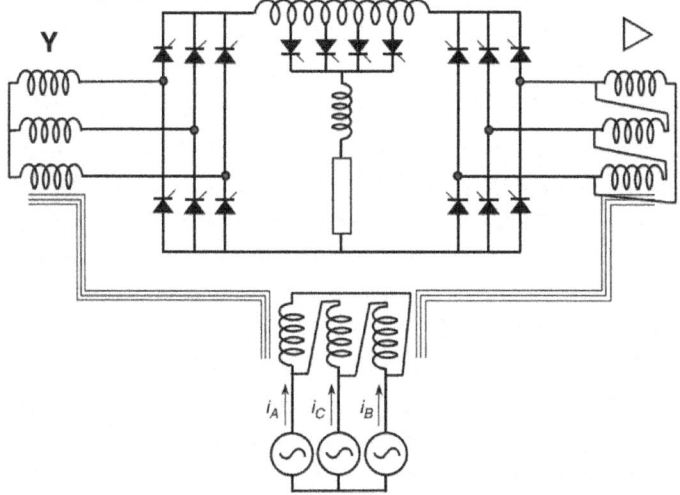

Figure 24: DC ripple reinjection technique for 48-pulse operation.

Higher pulse configuration using the same principle is also possible. The 12-pulse rectifier was obtained with a 30° phase shift between the two secondary transformers. The addition of further, appropriately shifted, transformers in parallel provides the basis for increasing pulse configurations. For instance, 24-pulse operation is achieved by means of four transformers with 15° phase shift, and 48-pulse operation requires eight transformers with 7.5° phase shift (transformer connections in zig-zag configuration).

Although theoretically possible, pulse numbers above 48 are rarely justified due to the practical levels of distortion found in the supply voltage waveforms. Further, the converter topology becomes more and more complicated.

An ingenious and very simple way to reach high pulse operation is shown in Fig. 24. This configuration is called dc ripple reinjection. It consists of two parallel converters connected to the load through a multistep reactor. The reactor uses a chain of thyristor-controlled taps, which are connected to symmetrical points of the reactor. By firing the thyristors located at the reactor at the right time, high-pulse operation is reached. The level of pulse operation depends on the number of thyristors connected to the reactor. They multiply thebasic level of operation of the two converters. The example, is Fig. 24, shows a 48-pulse configuration, obtained by the multiplication of basic 12-pulse operation by four reactor thyristors. This technique also can be applied to series connected bridges.

Another solution for harmonic reduction is the utilization of active power filters. Active power filters are special pulse width modulated (PWM) converters, able to generate the harmonics the converter requires. Figure 25 shows a current controlled shunt active power filter.

Applications of Line-commutated Rectifiers in Machine Drives

Important applications for line-commutated three-phase controlled rectifiers, are found in machine drives. Figure 26 shows a dc machine control implemented with a six-pulse rectifier. Torque and speed are controlled through the armature current I_D, and excitation current I_{exc}. Current I_D is adjusted with V_D, which is controlled by the firing angle α through Eq. (12). This dc drive can operate in two quadrants: positive and negative dc voltage. This two-quadrant operation allows regenerative braking when $\alpha > 90°$ and $I_{exc} < 0$.

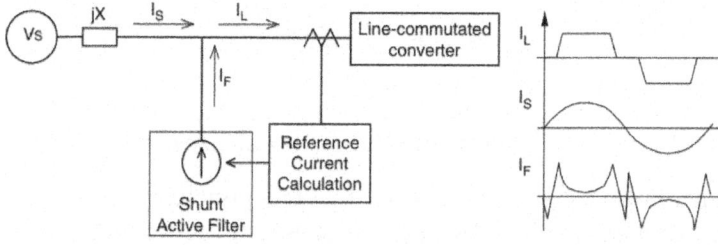

Figure 25: Current controlled shunt active power filter.

The converter of Fig. 26 can also be used to control synchronous machines, as shown in Fig. 27. In this case, a second converter working in the inverting mode operates the machine as self-controlled synchronous motor. With this second converter, the synchronous motor behaves like a dc motor but has none of the disadvantages of mechanical commutation. This converter is not line commutated, but machine commutated.

The nominal synchronous speed of the motor on a 50 or 60 Hz ac supply is now meaningless and the upper speed limit is determined by the mechanical limitations of the rotor construction. There is disadvantage that the rotational emfs required for load commutation of the machine side converter are not available at standstill and low speeds. In such a case, auxiliary force commutated circuits must be used.

The line-commutated rectifier controls the torque of the machine through firing angle α. This approach gives direct torque control of the commutatorless motor and is analogous to the use of armature current control shown in Fig. 26 for the converter-fed dc motor drive.

Line-commutated rectifiers are also used for speed control of wound rotor induction motors. Subsynchronous and supersynchronous static converter cascades using a naturally commutated dc link converter, can be implemented. Figure 28 shows a supersynchronous cascade for a wound rotor induction motor, using a naturally commutated dc link converter.

In the supersynchronous cascade shown in Fig. 28, the right hand bridge operates at slip frequency as a rectifier or inverter, while the other operates at network frequency as an inverter or rectifier. Control is difficult near synchronism when slip frequency emfs are insufficient for natural commutation and special circuit configuration employing forced commutation or devices with a self-turn-off capability are necessary for the passage through synchronism. This kind of supersynchronous cascade works better with cycloconverters.

Applications in HVDC Power Transmission

High voltage direct current (HVDC) power transmission is the most powerful application for line-commutated converters that exist today. There are power converters with ratings in excess of 1000 MW. Series operation of hundreds of valves can be found in some HVDC systems. In high power and long distance applications, these systems become more economical than conventional ac systems. They also have some other advantages compared with ac systems:

1. they can link two ac systems operating unsynchronized or with different nominal frequencies, that is 50 ↔ 60 Hz;

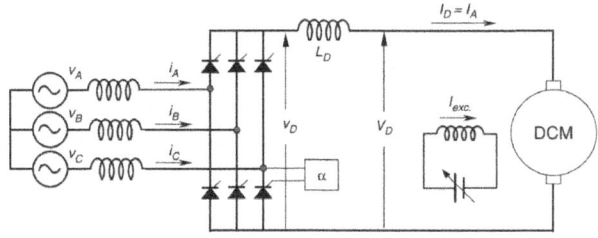

Figure 26: DC machine drive with a six-pulse rectifier.

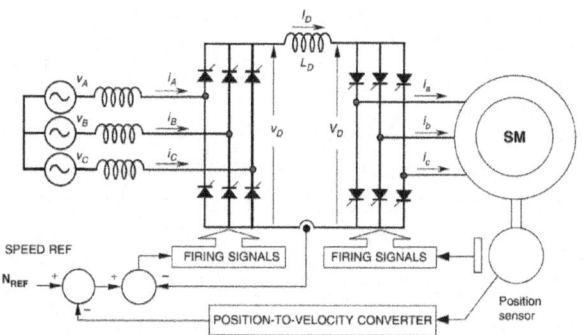

Figure 27: Self-controlled synchronous motor drive.

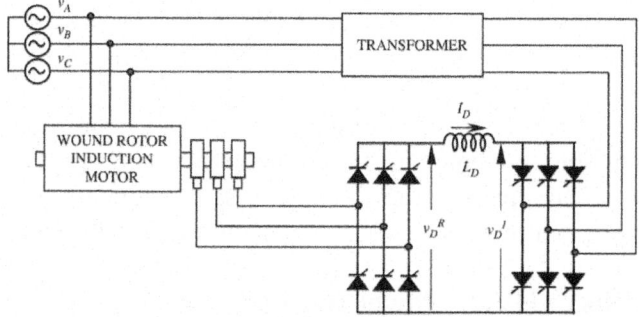

Figure 28: Supersynchronous cascade for a wound rotor induction motor.

- they can help in stability problems related with subsynchronous resonance in long ac lines;

- they have very good dynamic behavior and can interrupt short-circuit problems very quickly;

- if transmission is by submarine or underground cable, it is not practical to consider ac cable systems exceeding 50 km, but dc cable transmission systems are in service whose length is in hundreds of kilometers and even distances of 600 km or greater have been considered feasible;

- reversal of power can be controlled electronically by means of the delay firing angles α; and

- some existing overhead ac transmission lines cannot be increased. If overbuilt with or upgraded to dc transmission can substantially increase the power transfer capability on the existing right-of-way.

The use of HVDC systems for interconnections of asynchronous systems is an interesting application. Some continental electric power systems consist of asynchronous networks such as those for the East, West, Texas, and Quebec networks in North America, and islands loads such as that for the Island of Gotland in the Baltic Sea make good use of the HVDC interconnections.

Nearly all HVDC power converters with thyristor valves are assembled in a converter bridge of 12-pulse configuration, as shown in Fig. 29. Consequently, the ac voltages applied to each six-pulse valve group which makes up the 12-pulse valve group have a phase difference of 30° which is utilized to cancel the ac side, 5th and 7th harmonic currents and dc side, 6th harmonic voltage, thus resulting in a significant saving in harmonic filters.

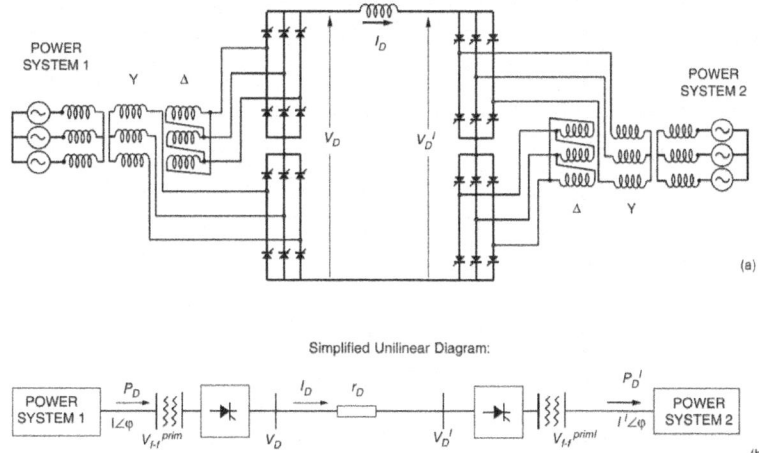

Figure 29: Typical HVDC power system: (a) detailed circuit and (b) unilinear diagram.

Some useful relations for HVDC systems include:

Rectifier Side:

$$P_D = V_D \cdot I_D = \sqrt{3} \cdot V_{f-f}^{prim} \cdot I_{line}^{rms} \cos \varphi \qquad (40)$$

$$I_P = I \cos \varphi$$

$$I_Q = I \sin \varphi$$

$$\therefore P_D = V_D \cdot I_D = \sqrt{3} \cdot V_{f-f}^{prim} \cdot I_P \tag{41}$$

$$I_P = \frac{V_D \cdot I_D}{\sqrt{3} \cdot V_{f-f}^{prim}} \tag{42}$$

$$I_P = \frac{a^2 \sqrt{3} \cdot V_{f-f}^{prim}}{4\pi \cdot \omega L_S} [\cos 2\alpha - \cos 2(\alpha + \mu)] \tag{43}$$

$$I_Q = \frac{a^2 \sqrt{3} \cdot V_{f-f}^{prim}}{4\pi \cdot \omega L_S} [\sin 2(\alpha + \mu) - \sin 2\alpha - 2\mu] \tag{44}$$

$$I_P = I_D \frac{a\sqrt{6}}{\pi} \left[\frac{\cos \alpha + \cos(\alpha + \mu)}{2} \right] \tag{45}$$

Fundamental secondary component of I

$$I = \frac{a\sqrt{6}}{\pi} I_D \tag{46}$$

Replacing Eq. (46) in (45)

$$I_P = I \cdot \left[\frac{\cos \alpha + \cos(\alpha + \mu)}{2} \right] \tag{47}$$

as $I_p = I \cos \varphi$, it yields Fig. 30a

$$\cos \varphi = \left[\frac{\cos \alpha + \cos(\alpha + \mu)}{2} \right] \tag{48}$$

Inverter Side:

The same equations are applied for the inverter side, but the firing angle α is replaced by γ, where γ is

$$\gamma = 180° - (\alpha + \mu) \tag{49}$$

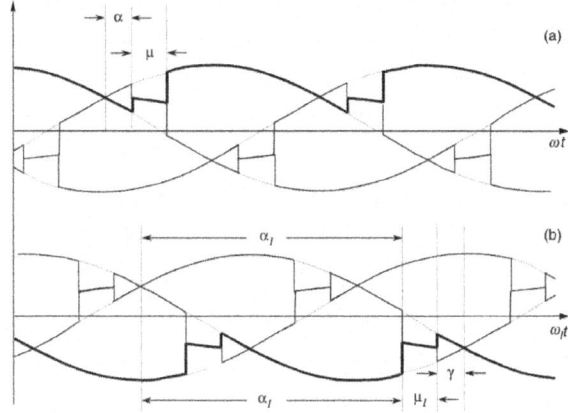

Figure 30: Definition of angle γ for inverter side: (a) rectifier side and (b) inverter side.

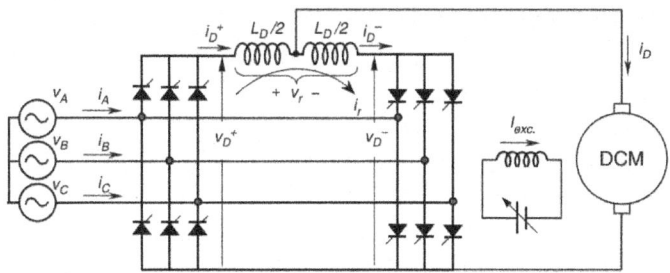

Figure 31: Dual converter in a four-quadrant dc drive.

As reactive power always goes in the converter direction, Eq. (44) for inverter side becomes (Fig. 30b)

$$I_{QI} = -\frac{a_I^2 \sqrt{3} \cdot V_{f-f_I}^{prim}}{4\pi \cdot \omega_I L_I} \left[\sin 2(\gamma + \mu_I) - \sin 2\gamma - 2\mu_I\right] \tag{50}$$

Dual Converters

In many variable-speed drives, four-quadrant operation is required, and three-phase dual converters are extensively used in applications up to 2 MW level. Figure 31 shows a threephase dual converter, where two converters are connected back-to-back.

In the dual converter, one rectifier provides the positive current to the load and the other the negative current. Due to the instantaneous voltage differences between the output voltages of the converters, a circulating current flows through the bridges. The circulating current is normally limited by circulating reactor, L_D, as shown in Fig. 31. The two converters are controlled in such a way that if α^+ is the delay angle of the positive current converter, the delay angle of the negative current converter is $\alpha^- = 180° - \alpha^+$.

Figure 32 shows the instantaneous dc voltages of each converter, v_D^+ and v_D^-. Despite the average voltage V_D is the same in both the converters, their instantaneous voltage differences shown as voltage v_r, are producing the circulating current i_r, which is superimposed with the load currents i_D^+ and i_D^-.

To avoid the circulating current i_r, it is possible to implement a "circulating current free" converter if a dead time of a few milliseconds is acceptable. The converter section, not required to supply current, remains fully blocked. When a current reversal is required, a logic switch-over system determines at first the instant at which the conducting converter's current becomes zero.

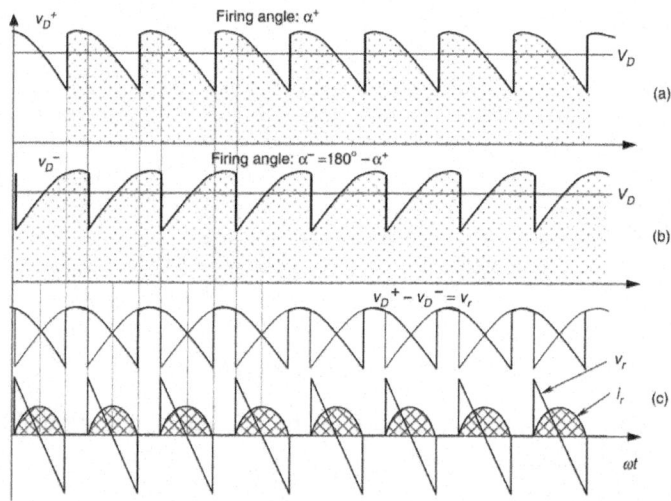

Figure 32: Waveform of circulating current: (a) instantaneous dc voltage from positive converter: (b) instantaneous dc voltage from negative converter; and (c) voltage difference between v_D^+ and v_D^-, v_r, and circulating current i_r.

This converter section is then blocked and the further supply of gating pulses to it is prevented. After a short safety interval (dead time), the gating pulses for the other converter section are released.

Cycloconverters

A different principle of frequency conversion is derived from the fact that a dual converter is able to supply an ac load with a lower frequency than the system frequency. If the control signal of the dual converter is a function of time, the output voltage will follow this signal. If this control signal value alters sinusoidally with the desired frequency, then the waveform depicted in Fig. 33a consists of a single-phase voltage with a large harmonic current. As shown in Fig. 33b, if the load is inductive, the current will present less distortion than voltage.

The cycloconverter operates in all four quadrants during a period. A pause (dead time) at least as small as the time required by the switch-over logic occurs after the current reaches zero, that is, between the transfer to operation in the quadrant corresponding to the other direction of current flow. Three single-phase cycloconverters may be combined to build a three-phase cycloconverter. The three-phase cycloconverters find an application in low-frequency, high-power requirements. Control speed of large synchronous motors in the low-speed range is one of the most common applications of three-phase cycloconverters. Figure 34 is a diagram for this application. They are also used to control slip frequency in wound rotor induction machines, for supersynchronous cascade (Scherbius system).

Harmonic Standards and Recommended Practices

In view of the proliferation of the power converter equipment connected to the utility system, various national and international agencies have been considering limits on harmonic current injection to maintain good power quality. As a consequence, various standards and guidelines have been established that specify limits on the magnitudes of harmonic currents and harmonic voltages.

The Comité Européen de Normalisation Electrotechnique (CENELEC), International Electrical Commission (IEC), and West German Standards (VDE) specify the limits on the voltages (as a percentage of the nominal voltage) at various harmonics frequencies of the utility frequency, when the equipment-generated harmonic currents are injected into a network whose impedances are specified.

According with Institute of Electrical and Electronic Engineers-519 standards (IEEE), Table 1 lists the limits on the harmonic currents that a user of power electronics equipment and other non-linear loads is allowed to inject into the utility system. Table 2 lists the quality of voltage that the utility can furnish the user.

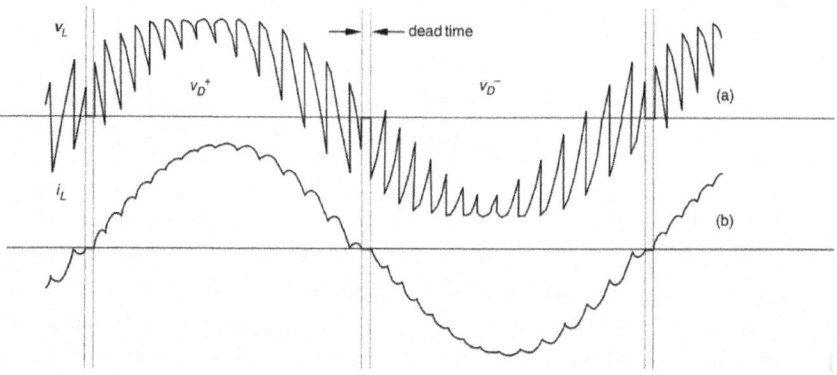

Figure 33: Cycloconverter operation: (a) voltage waveform and (b) current waveform for inductive load.

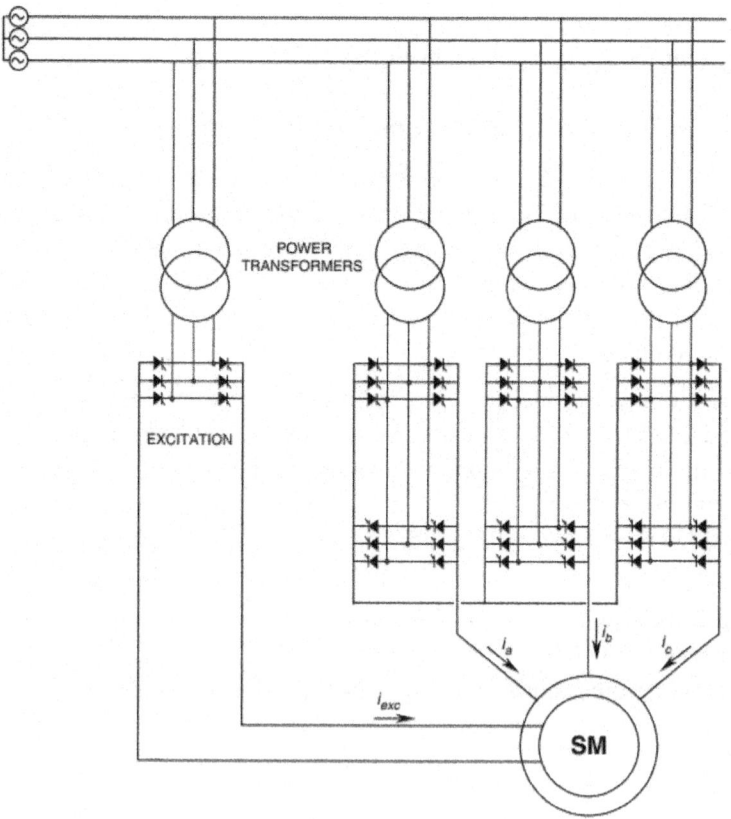

Figure 34: Synchronous machine drive with a cycloconverter.

Table 1: Harmonic current limits in percent of fundamental

Short circuit current (pu)	$h < 11$	$11 < h < 17$	$17 < h < 23$	$23 < h < 35$	$35 < h$	THD
<20	4.0	2.0	1.5	0.6	0.3	5.0
20–50	7.0	3.5	2.5	1.0	0.5	8.0
50–100	10.0	4.5	4.0	1.5	0.7	12.0
100–1000	12.0	5.5	5.0	2.0	1.0	15.0
>1000	15.0	7.0	6.0	2.5	1.4	20.0

Table 2: Harmonic voltage limits in percent of fundamental

Voltage level	2.3–69 kV	69–138 kV	>138 kV
Maximum for individual harmonic	3.0	1.5	1.0
Total harmonic distortion (THD)	5.0	2.5	1.5

In Table 1, the values are given at the point of connection of non-linear loads. The THD is the total harmonic distortion given by Eq. (51) and h is the number of the harmonic.

$$THD = \frac{\sqrt{\sum_{h=2}^{\infty} I_h^2}}{I_1} \qquad (51)$$

The total current harmonic distortion allowed in Table 1 increases with the value of short circuit current.

The total harmonic distortion in the voltage can be calculated in a manner similar to that given by Eq. (51). Table 2 specifies the individual harmonics and the THD limits on the voltage that the utility supplies to the user at the connection point.

FORCE-COMMUTATED THREE-PHASE CONTROLLED RECTIFIERS

Basic Topologies and Characteristics

Force-commutated rectifiers are built with semiconductors with gate-turn-off capability. The gate-turn-off capability allows full control of the converter, because valves can be switched ON and OFF whenever is required. This allows the commutation of the valves, hundreds of times in one period which is not possible with line-commutated rectifiers, where thyristors are switched ON and OFF only once a cycle. This feature has the following advantages:

(a) the current or voltage can be modulated (PWM), generating less harmonic contamination; (b) power factor can be controlled and even it can be made leading; and (c) they can be built as voltage source or current source rectifiers; (d) the reversal of power in thyristor rectifiers is by reversal of voltage at the dc link. Instead, force-commutated rectifiers can be implemented for both, reversal of voltage or reversal of current.

There are two ways to implement force-commutated threephase rectifiers: (a) as a current source rectifier, where power reversal is by dc voltage reversal; and (b) as a voltage source rectifier, where power reversal is by current reversal at the dc link. Figure 35 shows the basic circuits for these two topologies.

Operation of the Voltage Source Rectifier

The voltage source rectifier is by far the most widely used, and because of the duality of the two topologies showed in Fig. 35, only this type of force-commutated rectifier will be explained in detail.

The voltage source rectifier operates by keeping the dc link voltage at a desired reference value, using a feedback control loop as shown in Fig. 36. To accomplish this task, the dc link voltage is measured and compared with a reference VREF . The error signal generated from this comparison is used to switch the six valves of the rectifier ON and OFF. In this way, power can come or return to the ac source according with the dc link voltage requirements. The voltage V_D is measured at the capacitor C_D.

When the current I_D is positive (rectifier operation), the capacitor C_D is discharged, and the error signal ask the control block for more power from the ac supply. The control block takes the power from the supply by generating the appropriate PWM signals for the six valves. In this way, more current flows from the ac to the dc side and the capacitor voltage is recovered. Inversely, when I_D becomes negative (inverter operation), the capacitor C_D is overcharged and the error signal ask the control to discharge the capacitor and return power to the ac mains.

The PWM control can manage not only the active power, but also the reactive power, allowing this type of rectifier to correct power factor. In addition, the ac current waveforms can be maintained as almost sinusoidal, which reduces harmonic contamination to the mains supply.

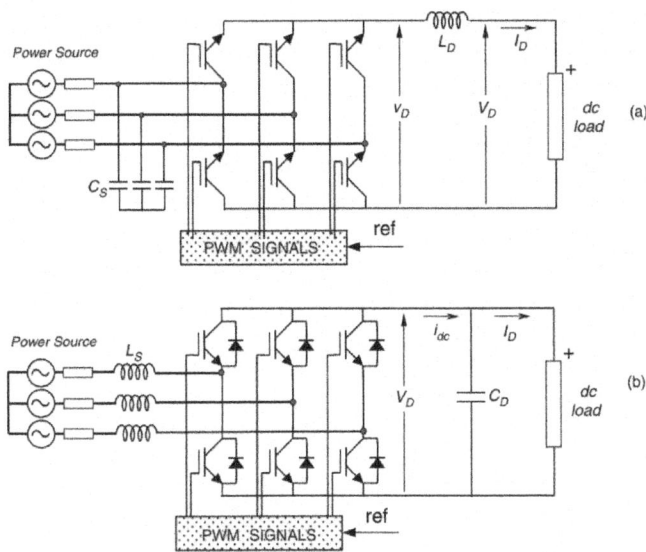

Figure 35: Basic topologies for force-commutated PWM rectifiers: (a) current source rectifier and (b) voltage source rectifier.

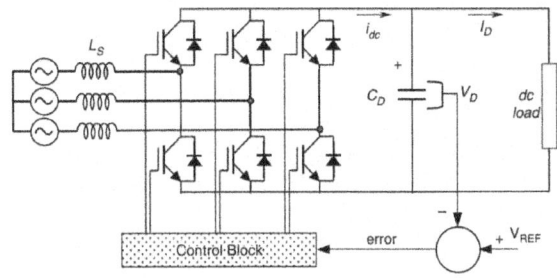

Figure 36: Operation principle of the voltage source rectifier.

The PWM consists of switching the valves ON and OFF, following a pre-established template. This template could be a sinusoidal waveform of voltage or current. For example, the modulation of one phase could be as the one shown in Fig. 37. This PWM pattern is a periodical waveform whose fundamental is a voltage with the same frequency of the template. The amplitude of this fundamental, called V_{MOD} in Fig. 37, is also proportional to the amplitude of the template.

To make the rectifier work properly, the PWM pattern must generate a fundamental V_{MOD} with the same frequency as the power source. Changing the amplitude of this fundamental and its phase shift with respect to the mains, the rectifier can be controlled to operate in the four quadrants: leading power factor rectifier, lagging power factor rectifier, leading power factor inverter, and lagging power factor inverter. Changing the pattern of modulation, as shown in Fig. 38, modifies the magnitude of V_{MOD}. Displacing the PWM pattern changes the phase shift.

The interaction between V_{MOD} and V (source voltage) can be seen through a phasor diagram. This interaction permits understanding of the four-quadrant capability of this rectifier. In Fig. 39, the following operations are displayed: (a) rectifier at unity power factor; (b) inverter at unity power factor; (c) capacitor (zero power factor); and (d) inductor (zero power factor).

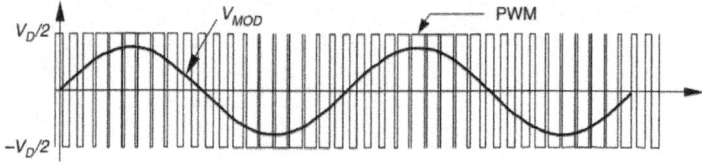

Figure 37: A PWM pattern and its fundamental V_{MOD}.

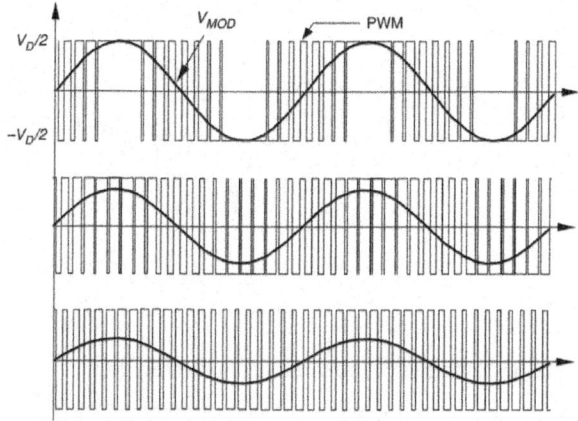

Figure 38: Changing VMOD through the PWM pattern.

In Fig. 39, I_S is the rms value of the source current is. This current flows through the semiconductors in the way shown in Fig. 40. During the positive half cycle, the transistor T_N , connected at the negative side of the dc link is switched ON, and the current is begins to flow through T_N (i_{Tn}). The current returns to the mains and comes back to the valves, closing a loop with another phase, and passing through a diode connected at the same negative terminal of the dc link. The current can also go to the dc load (inversion) and return through another transistor located at the positive terminal of the dc link. When the transistor T_N is switched OFF, the current path is interrupted and the current begins to flow through the diode D_p, connected at the positive terminal of the dc link. This current, called i_{Dp} in Fig. 39, goes directly to the dc link, helping in the generation of the current idc. The current idc charges the capacitor C_D and permits the rectifier to produce dc power. The inductances L_S are very important in this process, because they generate an induced voltage which allows the conduction of the diode DP . Similar operation occurs during the negative half cycle, but with T_p and D_N (see Fig. 40).

Under inverter operation, the current paths are different because the currents flowing through the transistors come mainly from the dc capacitor, C_D. Under rectifier operation, the circuit works like a boost converter and under inverter, it works as a buck converter.

To have full control of the operation of the rectifier, their six diodes must be polarized negatively at all values of instantaneous ac voltage supply. Otherwise diodes will conduct, and the PWM rectifier will behave like a common diode rectifier bridge. The way to keep the diodes blocked is to ensure a dc link voltage higher than the peak dc voltage generated by the diodes alone, as shown in Fig. 41. In this way, the diodes remain polarized negatively, and they will conduct only when at least one transistor is switched ON, and favorable instantaneous ac voltage conditions are given. In the Fig. 41, VD represents the capacitor dc voltage, which is kept higher than the normal diode-bridge rectification value v_{BRIDGE} . To maintain this condition, the rectifier must have a control loop like the one displayed in Fig. 36.

PWM Phase-to-Phase and Phase-to-Neutral Voltages

The PWM waveforms shown in the preceding figures are voltages measured between the middle point of the dc voltage and the corresponding phase. The phase-to-phase PWM voltages can be obtained with the help of Eq. (52), where the voltage V_{PWM}^{AB} is evaluated.

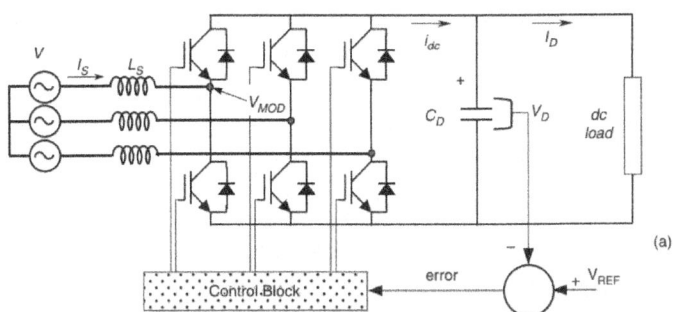

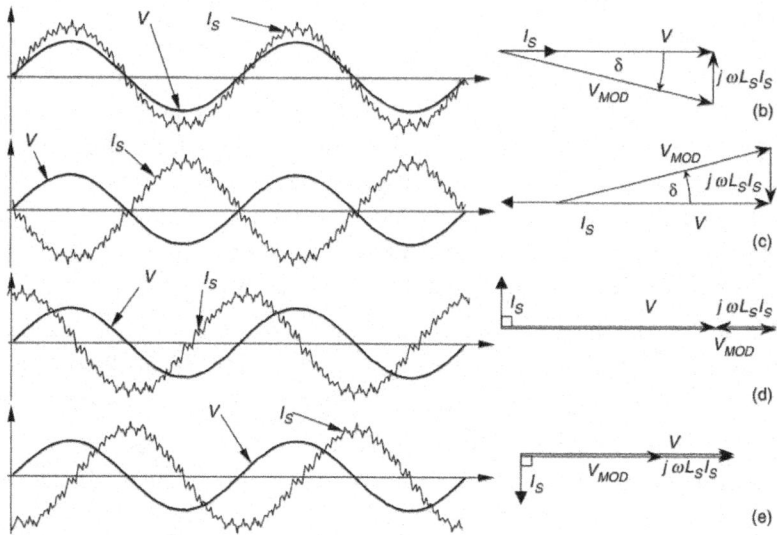

Figure 39: Four-quadrant operation of the force-commutated rectifier: (a) the PWM force-commutated rectifier; (b) rectifier operation at unity power factor; (c) inverter operation at unity power factor; (d) capacitor operation at zero power factor; and (e) inductor operation at zero power factor.

$$V_{PWM}^{AB} = V_{PWM}^{A} - V_{PWM}^{B} \tag{52}$$

where V_{PWM}^{A} and V_{PWM}^{B} are the voltages measured between the middle point of the dc voltage, and the phases a and b respectively. In a less straightforward fashion, the phase-toneutral voltage can be evaluated with the help of Eq. (53).

$$V_{PWM}^{AN} = \frac{1}{3}(V_{PWM}^{AB} - V_{PWM}^{CA}) \tag{53}$$

where V_{PWM}^{AN} is the phase-to-neutral voltage for phase a, and V_{PWM}^{jk} is the phase-to-phase voltage between phase j and phase k. Figure 42 shows the PWM patterns for the phase-to-phase and phase-to-neutral voltages.

Control of the DC Link Voltage

Control of the dc link voltage requires a feedback control loop. The dc voltage V_D is compared with a reference V_{REF}, and the error signal "e" obtained from this comparison is used to generate a template waveform. The template should be a sinusoidal waveform with the same frequency of the mains supply. This template is used to produce the PWM pattern and allows controlling the rectifier in two different ways: (1) as a voltage-source current-controlled PWM rectifier or (2) as a voltage-source voltage-controlled PWM rectifier.

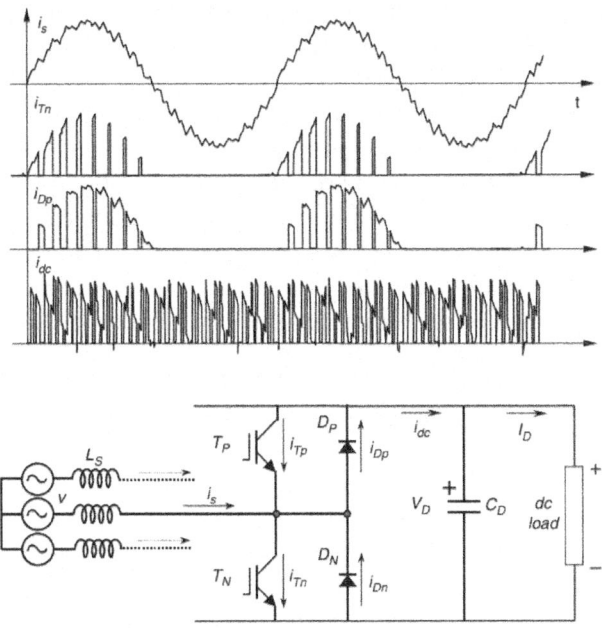

Figure 40: Current waveforms through the mains, the valves, and the dc link.

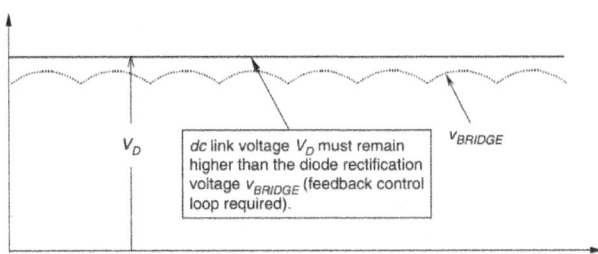

Figure 41: The DC link voltage condition for the operation of the PWM rectifier.

The first method controls the input current, and the second controls the magnitude and phase of the voltage V_{MOD}. The current-controlled method is simpler and more stable than the voltage-controlled method, and for these reasons it will be explained first.

Voltage-source Current-controlled PWM Rectifier

This method of control is shown in the rectifier in Fig. 43. Control is achieved by measuring the instantaneous phase currents and forcing them to follow a sinusoidal current reference template, I_ref. The amplitude of the current reference template, I_{MAX} is evaluated using the following equation

$$I_{MAX} = G_C \cdot e = G_C \cdot (V_{REF} - v_D)$$

(54)

where G_c is shown in Fig. 43 and represents a controller such as PI, P, Fuzzy, or other. The sinusoidal waveform of the template is obtained by multiplying I_{MAX} with a sine function, with the same frequency of the mains, and with the desired phase-shift angle φ, as shown in Fig. 43. Further, the template must be synchronized with the power supply. After that, the template has been created and is ready to produce the PWM pattern.

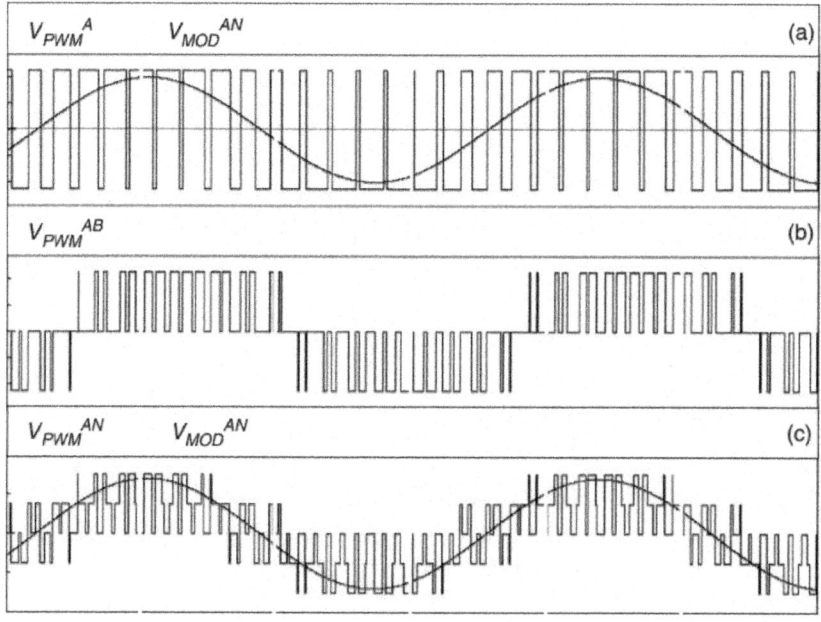

Figure 42: PWM phase voltages: (a) PWM phase modulation; (b) PWM phase-to-phase voltage; and (c) PWM phase-to-neutral voltage.

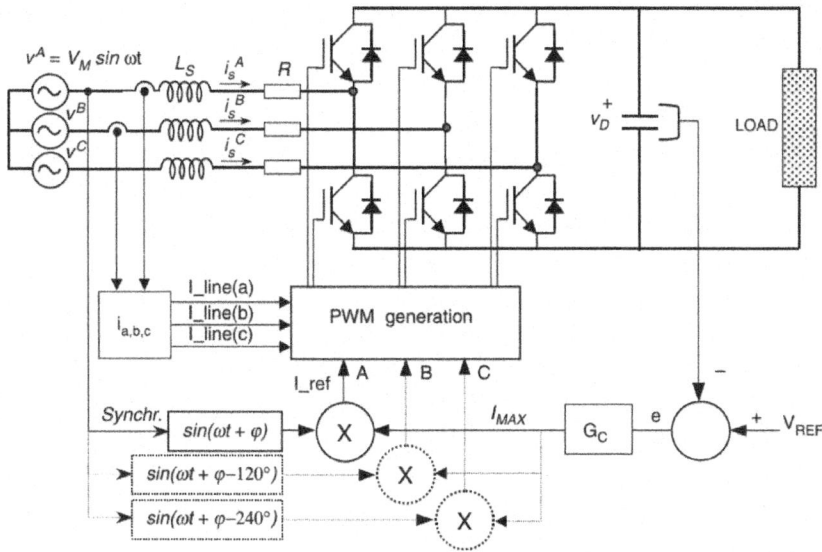

Figure 43: Voltage-source current-controlled PWM rectifier.

However, one problem arises with the rectifier because the feedback control loop on the voltage V_C can produce instability. Then it becomes necessary to analyze this problem during rectifier design. Upon introducing the voltage feedback and the G_C controller, the control of the rectifier can be represented in a block diagram in Laplace dominion, as shown in Fig. 44. This block diagram represents a linearization of the system around an operating point, given by the rms value of the input current, I_S.

The blocks $G_1(s)$ and $G_2(s)$ in Fig. 44 represent the transfer function of the rectifier (around the operating point) and the transfer function of the dc link capacitor C_D respectively.

$$G_1(s) = \frac{\Delta P_1(s)}{\Delta I_S(s)} = 3 \cdot (V \cos \varphi - 2RI_S - L_SI_Ss) \tag{55}$$

$$G_2(s) = \frac{\Delta V_D(s)}{\Delta P_1(s) - \Delta P_2(s)} = \frac{1}{V_D \cdot C_D \cdot s} \tag{56}$$

where $P_1(s)$ and $P_2(s)$ represent the input and output power of the rectifier in Laplace dominion, V the rms value of the mains voltage supply (phase-to-neutral), IS the input current being controlled by the template, L_S the input inductance, and R the resistance between the converter and power supply.

According to stability criteria, and assuming a PI controller, the following relations are obtained

$$I_S \leq \frac{C_D \cdot V_D}{3 K_P \cdot L_S} \qquad (57)$$

$$I_S \leq \frac{K_P \cdot V \cdot \cos \varphi}{2R \cdot K_P + L_S \cdot K_I} \qquad (58)$$

These two relations are useful for the design of the currentcontrolled rectifier. They relate the values of dc link capacitor, dc link voltage, rms voltage supply, input resistance and inductance, and input power factor, with the rms value of the input current, I_s . With these relations the proportional and integral gains K_p and K_I can be calculated to ensure the stability of the rectifier. These relations only establish limitations for rectifier operation, because negative currents always satisfy the inequalities.

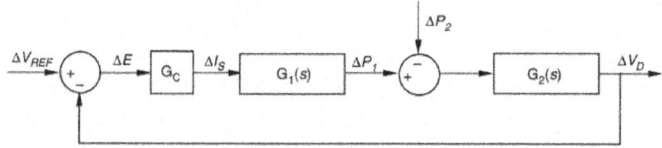

Figure 44: Closed-loop rectifier transfer function.

With these two stability limits satisfied, the rectifier will keep the dc capacitor voltage at the value of V_{REF} (PI controller), for all load conditions, by moving power from the ac to the dc side. Under inverter operation, the power will move in the opposite direction.

Once the stability problems have been solved and the sinusoidal current template has been generated, a modulation method will be required to produce the PWM pattern for the power valves. The PWM pattern will switch the power valves to force the input currents I_line to follow the desired current template I_ref. There are many modulation methods in the literature, but three methods for voltage-source current-controlled rectifiers are the most widely used ones: periodical sampling (PS), hysteresis band (HB), and triangular carrier (TC).

The PS method switches the power transistors of the rectifier during the transitions of a square wave clock of fixed frequency: the periodical sampling frequency. In each transition, a comparison between I_ref and I_line is made, and corrections take place. As shown in Fig. 45a, this type of control is very simple to implement: only a comparator and a D-type flip-flop are needed per phase. The main advantage of this method is that the minimum time between switching transitions is limited to the period of the sampling clock.

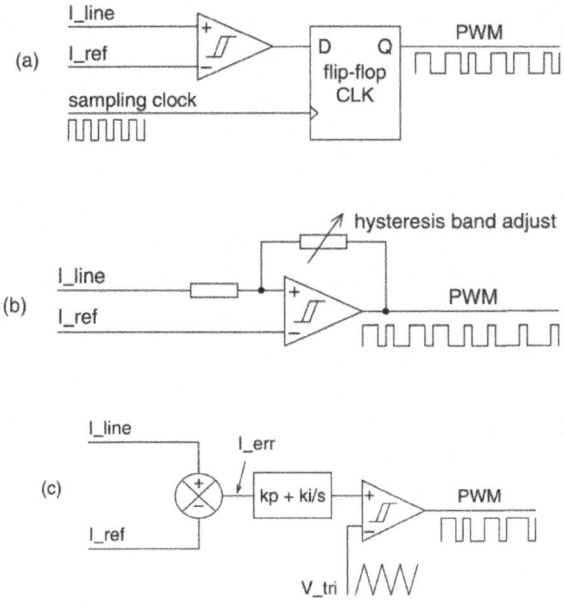

Figure 45: Modulation control methods: (a) periodical sampling; (b) hysteresis band; and (c) triangular carrier.

This characteristic determines the maximum switching frequency of the converter. However, the average switching frequency is not clearly defined.

The HB method switches the transistors when the error between I_ref and I_line exceeds a fixed magnitude: the hysteresis band. As it can be seen in Fig. 45b, this type of control needs a single comparator with hysteresis per phase. In this case the switching frequency is not determined, but its maximum value can be evaluated through the following equation

$$f_S^{max} = \frac{V_D}{4h \cdot L_S}$$

(59)

where h is the magnitude of the hysteresis band.

The TC method, shown in Fig. 45c, compares the error between I_ref and I_line with a triangular wave. This triangular wave has fixed amplitude and frequency and is called the triangular carrier. The error is processed through a proportional-integral (PI) gain stage before comparison with the TC takes place. As can be seen, this control scheme is more complex than PS and HB.

The values for k_p and k_i determine the transient response and steady-state error of the TC method. It has been found empirically that the values for k_p and k_i shown in Eqs. (60) and (61) give a good dynamic performance under several operating conditions.

$$k_p = \frac{L_s \cdot \omega_c}{2 \cdot V_D}$$

(60)

$$k_i = \omega_c \cdot K_P$$

(61)

where L_S is the total series inductance seen by the rectifier, ω_c is the T_C frequency, and V_D is the dc link voltage of the rectifier.

In order to measure the level of distortion (or undesired harmonic generation) introduced by these three control methods, Eq. (62) is defined

$$\%Distortion = \frac{100}{I_{rms}} \sqrt{\frac{1}{T} \int_T (I\_line - I\_ref)^2 \, dt}$$

(62)

In Eq. (62), the term I_{rms} is the effective value of the desired current. The term inside the square root gives the rms value of the error current, which is undesired. This formula measures the percentage of error (or distortion) of the generated waveform. This definition considers the ripple, amplitude, and phase errors of the measured waveform, as opposed to the THD, which does not take into account offsets, scalings, and phase shifts.

Figure 46 shows the current waveforms generated by the three aforementioned methods. The example uses an average switching frequency of 1.5 kHz. The PS is the worst, but its digital implementation is simpler. The HB method and TC with PI control are quite similar, and the TC with only proportional control gives a current with a small phase shift.

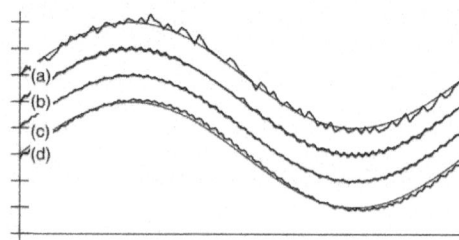

Figure 46: Waveforms obtained using 1.5 kHz switching frequency and $L_S = 13$ mH: (a) P_S method; (b) HB method; (c) T_C method ($K_P + K_I$); and (d) T_C method (K_P only).

However, Fig. 47 shows that the higher the switching frequency, the closer the results obtained with the different modulation methods. Over 6 kHz of switching frequency, the distortion is very small for all methods.

Voltage-source Voltage-controlled PWM Rectifier

Figure 48 shows a one-phase diagram from which the control system for a voltage-source voltage-controlled rectifier is derived. This diagram represents an equivalent circuit of the fundamentals, that is, pure sinusoidal at the mains side and pure dc at the dc link side. The control is achieved by creating a sinusoidal voltage template V_{MOD}, which is modified in amplitude and angle to interact with the mains voltage V. In this way the input currents are controlled without measuring them. The template V_{MOD} is generated using the differential equations that govern the rectifier.

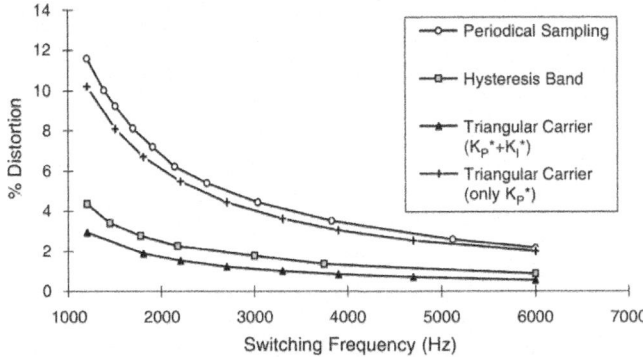

Figure 47: Distortion comparison for a sinusoidal current reference.

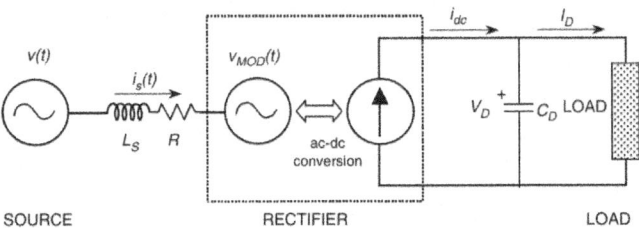

Figure 48: One-phase fundamental diagram of the voltage source rectifier.

The following differential equation can be derived from Fig. 48

$$v(t) = L_S \frac{di_s}{dt} + Ri_s + v_{MOD}(t)$$

$$(63)$$

Assuming that $v(t) = V\sqrt{2} \sin \omega t$, then the solution for $i_s(t)$, to acquire a template V_{MOD} able to make the rectifier work at constant power factor, should be of the form

$$i_s(t) = I_{max}(t) \sin(\omega t + \varphi)$$

$$(64)$$

Equations (63), (64), and v(t) allows a function of time able to modify V_{MOD} in amplitude and phase that will make the rectifier work at a fixed power factor. Combining these equations with v(t) yields

$$v_{MOD}(t) = \left[V\sqrt{2} + X_S I_{max} \sin\varphi - \left(RI_{max} + L_S \frac{dI_{max}}{dt} \right) \cos\varphi \right] \sin \omega t$$

$$- \left[X_S I_{max} \cos\varphi + \left(RI_{max} + L_S \frac{dI_{max}}{dt} \right) \sin\varphi \right] \cos \omega t$$

$$(65)$$

Equation (65) provides a template for V_{MOD}, which is controlled through variations of the input current amplitude I_{max}. Substituting the derivatives of I_{max} into Eq. (65) make sense, because I_{max} changes every time the dc load is modified. The term X_S in Eq. (65) is ωL_S. This equation can also be written for unity power factor operation. In such a case, $\cos \varphi = 1$ and $\sin \varphi = 0$.

$$v_{MOD}(t) = \left(V\sqrt{2} - RI_{max} - L_S \frac{dI_{max}}{dt} \right) \sin \omega t$$

$$- X_S I_{max} \cos \omega t$$

$$(66)$$

With this last equation, a unity power factor, voltage source, voltage-controlled PWM rectifier can be implemented as shown in Fig. 49. It can be observed that Eqs. (65) and (66) have an in-phase term with the mains supply (sin ωt) and an in-quadrature term (cos ωt). These two terms allow the template V_{MOD} to change in magnitude and phase so as to have full unity power factor control of the rectifier.

Compared with the control block of Fig. 43, in the voltage-source voltage-controlled rectifier of Fig. 49, there is no need to sense the input currents. However, to ensure stability limits as good as the limits of the current-controlled rectifier, the blocks "–R–sL$_S$" and "–X$_S$" in Fig. 49, have to emulate and reproduce exactly the real values of R, X$_S$, and L$_S$ of the power circuit. However, these parameters do not remain constant, and this fact affects the stability of this system, making it less stable than the system showed in Fig. 43. In theory, if the impedance parameters are reproduced exactly, the stability limits of this rectifier are given by the same equations as used for the current-controlled rectifier seen in Fig. 43 (Eqs. (57) and (58)).

Under steady-state, I_{max} is constant, and Eq. (66) can be written in terms of phasor diagram, resulting in Eq. (67). As shown in Fig. 50, different operating conditions for the unity power factor rectifier can be displayed with this equation

$$\vec{V}_{MOD} = \vec{V} - R\vec{I}_S - jX_S\vec{I}_S \qquad (67)$$

With the sinusoidal template VMOD already created, a modulation method to commutate the transistors will be required. As in the case of current-controlled rectifier, there are many methods to modulate the template, with the most well known the so-called sinusoidal pulse width modulation (SPWM), which uses a TC to generate the PWM as shown in Fig. 51. Only this method will be described in this chapter. In this method, there are two important parameters to define: the amplitude modulation ratio or modulation index m, and the frequency modulation ratio p. Definitions are given by

$$m = \frac{V_{MOD}^{max}}{V_{TRIANG}^{max}} \qquad (68)$$

$$p = \frac{f_T}{f_S} \qquad (69)$$

where V_{MOD}^{max} and V_{TRIANG}^{max} are the amplitudes of V_{MOD} and V_{TRIANG} respectively. On the other hand, f_S is the frequency of the mains supply and f_T the frequency of the TC. In Fig. 51, m = 0.8 and p = 21. When m > 1 overmodulation is defined.

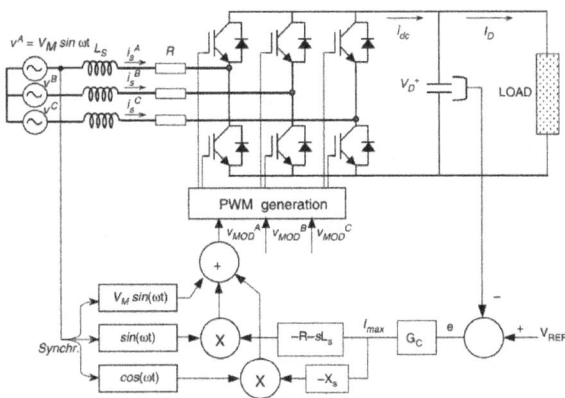

Figure 49: Implementation of the voltage-controlled rectifier for unity power factor operation.

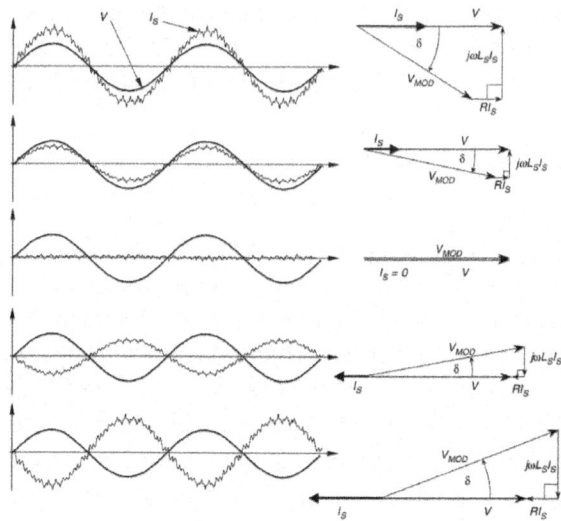

Figure 50: Steady-state operation of the unity power factor rectifier under different load conditions.

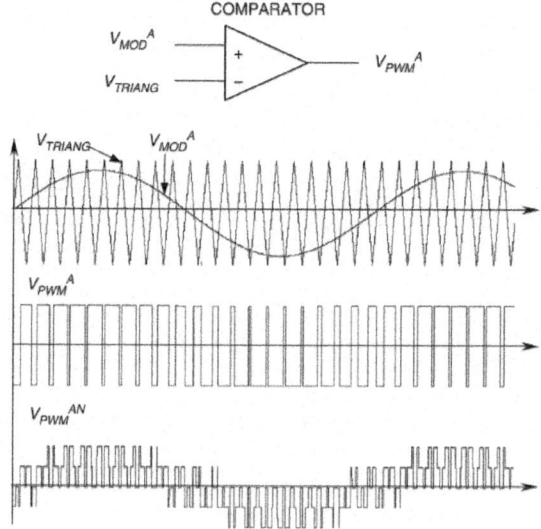

Figure 51: Sinusoidal modulation method based on TC.

The modulation method described in Fig. 51 has a harmonic content that changes with p and m. When p < 21, it is recommended that synchronous PWM be used, which means that the TC and the template should be synchronized. Furthermore, to avoid subharmonics, it is also desired that p be an integer. If p is an odd number, even harmonics will be eliminated. If p is a multiple of 3, then the PWM modulation of the three phases will be identical. When m increases, the amplitude of the fundamental voltage increases proportionally, but some harmonics decrease. Under overmodulation (m > 1), the fundamental voltage does not increase linearly, and more harmonics appear. Figure 52 shows the harmonic spectrum of the three-phase PWM voltage waveforms for different values of m, and p = 3k where k is an odd number.

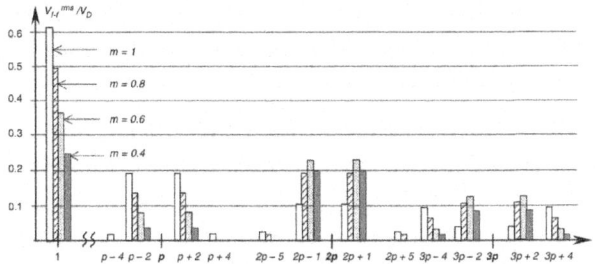

Figure 52: Harmonic spectrum for SPWM modulation.

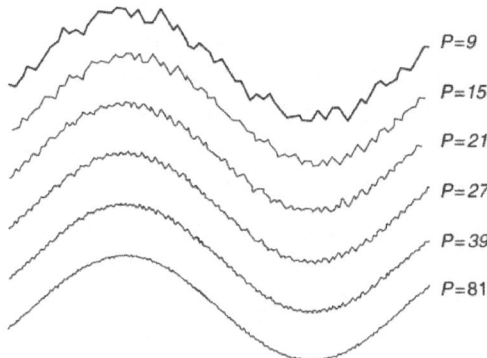

Figure 53: Current waveforms for different values of p.

Due to the presence of the input inductance L_S, the harmonic currents that result are proportionally attenuated with the harmonic number. This characteristic is shown in the current waveforms of Fig. 53, where larger p numbers generate cleaner currents. The rectifier that originated the currents of Fig. 53 has the following characteristics: V_D = 450 Vd$_c$, V_{f-f}^{rms} = 220 V$_{ac}$, L$_S$ = 2 mH, and input current I$_S$ = 80 Arms. It can be observed that with p > 21 the current distortion is quite small. The value of p = 81 in Fig. 53 produces an almost pure sinusoidal waveform, and it means 4860 Hz of switching frequency at 60 Hz or only 4.050 Hz in a rectifier operating in a 50 Hz supply.

This switching frequency can be managed by MOSFETs, IGBTs, and even Power Darlingtons. Then a number p = 81, is feasible for today's low and medium power rectifiers.

Voltage-source Load-controlled PWM Rectifier

A simple method of control for small PWM rectifiers (up to 10–20 kW) is based on direct control of the dc current. Figure 54 shows the schematic of this control system. The fundamental voltage V_{MOD} modulated by the rectifier is produced by a fixed and unique PWM pattern, which can be carefully selected to eliminate most undesirable harmonics. As the PWM does not change, it can be stored in a permanent digital memory (ROM).

The control is based on changing the power angle δ between the mains voltage V and fundamental PWM voltage V_{MOD}. When δ changes, the amount of power flow transferred from the ac to the dc side also changes. When the power angle is negative (V_{MOD} lags V), the power flow goes from the ac to the dc side. When the power angle is positive, the power flows in the opposite direction. Then, the power angle can be controlled through the current I_D. The voltage V_D does not need to be sensed, because this control establishes a stable dc voltage operation for each dc current and power angle. With these characteristics, it is possible to find a relation between I_D and δ so as to obtain constant dc voltage for all load conditions.

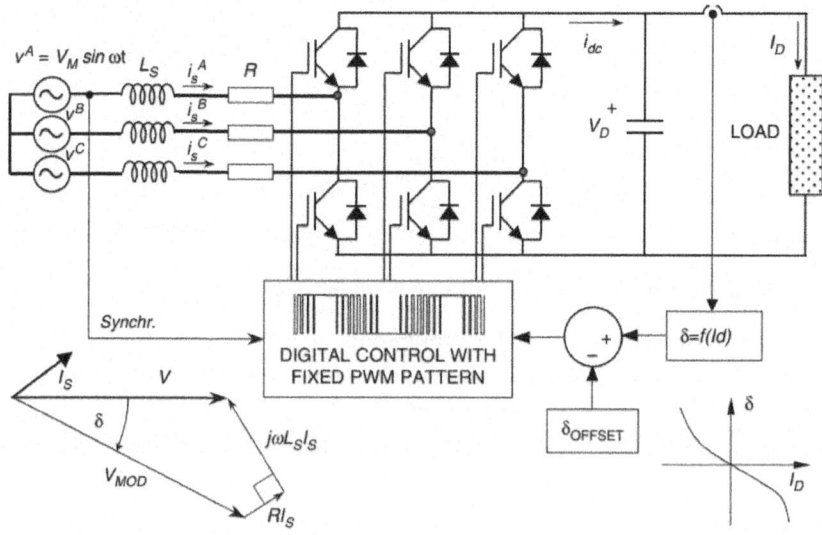

Figure 54: Voltage-source load-controlled PWM rectifier.

This relation is given by

$$I_D = f(\delta) = \frac{V\left(\cos\delta - \frac{\omega L_S}{R}\sin\delta - 1\right)}{R\left[1 + \left(\frac{\omega L_S}{R}\right)^2\right]}$$

(70)

From Eq. (70) a plot and a reciprocal function $\delta = f(I_D)$ is obtained to control the rectifier. The relation between I_D and δ allows for leading power factor operation and null regulation. The leading power factor operation is shown in the phasor diagram of Fig. 54.

The control scheme of the voltage-source load-controlled rectifier is characterized by the following: (i) there are neither input current sensors nor dc voltage sensor; (ii) it works with a fixed and predefined PWM pattern; (iii) it presents very good stability; (iv) its stability does not depend on the size of the dc capacitor; (v) it can work at leading power factor for all load conditions; and (vi) it can be adjusted with Eq. (70) to work at zero regulation. The drawback appears when R in Eq. (70) becomes negligible, because in such a case the control system is unable to find an equilibrium point for the dc link voltage. That is why this control method is not applicable to large systems.

New Technologies and Applications of Force-commutated Rectifiers

The additional advantages of force-commutated rectifiers with respect to line-commutated rectifiers, make them better candidates for industrial requirements. They permit new applications such as rectifiers with harmonic elimination capability (active filters), power factor compensators, machine drives with four-quadrant operation, frequency links to connect 50 Hz with 60 Hz systems, and regenerative converters for traction power supplies. Modulation with very fast valves such as IGBTs permit almost sinusoidal currents to be obtained. The dynamics of these rectifiers is so fast that they can reverse power almost instantaneously. In machine drives, current source PWM rectifiers, like the one shown in Fig. 35a, can be used to drive dc machines from the three-phase supply. Four-quadrant applications using voltage-source PWM rectifiers, are extended for induction machines, synchronous machines with starting control, and special machines such as brushless-dc motors. Back-to-back systems are being used in Japan to link power systems with different frequencies.

Active Power Filter

Force-commutated PWM rectifiers can work as active power filters. The voltage-source current-controlled rectifier has the capability to eliminate harmonics produced by other polluting loads. It only needs to be connected as shown in Fig. 55.

The current sensors are located at the input terminals of the power source and these currents (instead of the rectifier currents) are forced to be sinusoidal. As there are polluting loads in the system, the rectifier is forced to deliver the harmonics that loads need, because the current sensors do not allow the harmonics going to the mains. As a result, the rectifier currents become distorted, but an adequate dc capacitor C_D can keep the dc link voltage in good shape. In this way the rectifier can do its duty, and also eliminate harmonics to the source. In addition, it also can compensate power factor and unbalanced load problems.

Frequency Link Systems

Frequency link systems permit power to be transferred form one frequency to another one. They are also useful for linking unsynchronized networks. Line-commutated converters are widely used for this application, but they have some drawbacks that force-commutated converters can eliminate. For example, the harmonic filters requirement, the poor power factor, and the necessity to count with a synchronous compensator when generating machines at the load side are absent. Figure 56 shows a typical line-commutated system in which a 60 Hz load is fed by a 50 Hz supply. As the 60 Hz side needs excitation to commutate the valves, a synchronous compensator has been required.

In contrast, an equivalent system with force-commutated converters is simpler, cleaner, and more reliable. It is implemented with a dc voltage-controlled rectifier, and another identical converter working in the inversion mode. The power factor can be adjusted independently at the two ac terminals, and filters or synchronous compensators are not required. Figure 57 shows a frequency link system with force-commutated converters.

Special Topologies for High Power Applications

High power applications require series- and/or parallelconnected rectifiers. Series and parallel operation with forcecommutated rectifiers allow improving the power quality because harmonic cancellation can be applied to these topologies. Figure 58 shows a series connection of forcecommutated rectifiers, where the modulating carriers of the valves in each bridge are shifted to cancel harmonics. The example uses sinusoidal PWM that are with TC shifted.

The waveforms of the input currents for the series connection system are shown in Fig. 59. The frequency modulation ratio shown in this figure is for $p = 9$. The carriers are shifted by 90° each, to obtain harmonics cancellation.

Shifting of the carriers δ_T depends on the number of converters in series (or in parallel), and is given by

$$\delta_T = \frac{2\pi}{n} \qquad (71)$$

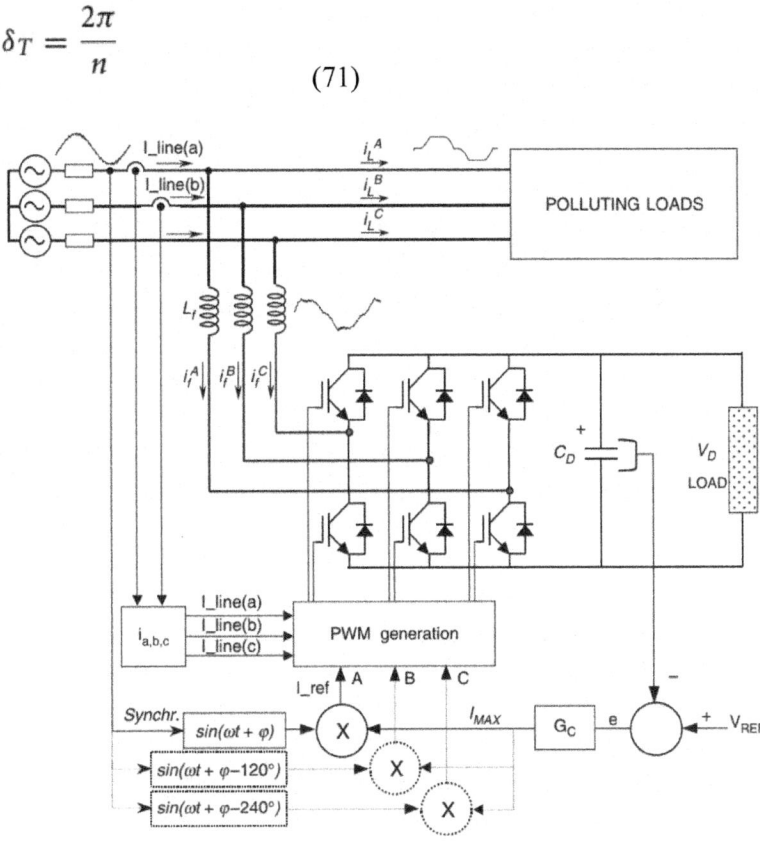

Figure 55: Voltage-source rectifier with harmonic elimination capability.

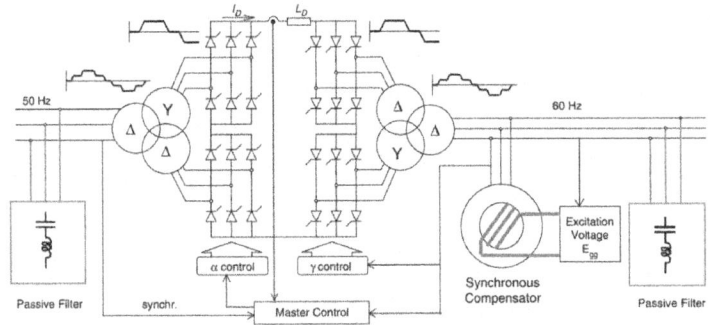

Figure 56: Frequency link systems with line-commutated converters.

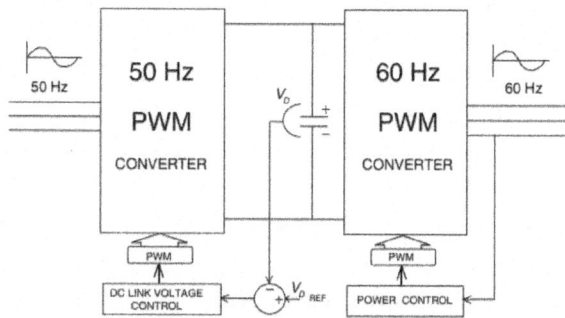

Figure 57: Frequency link systems with force-commutated converters.

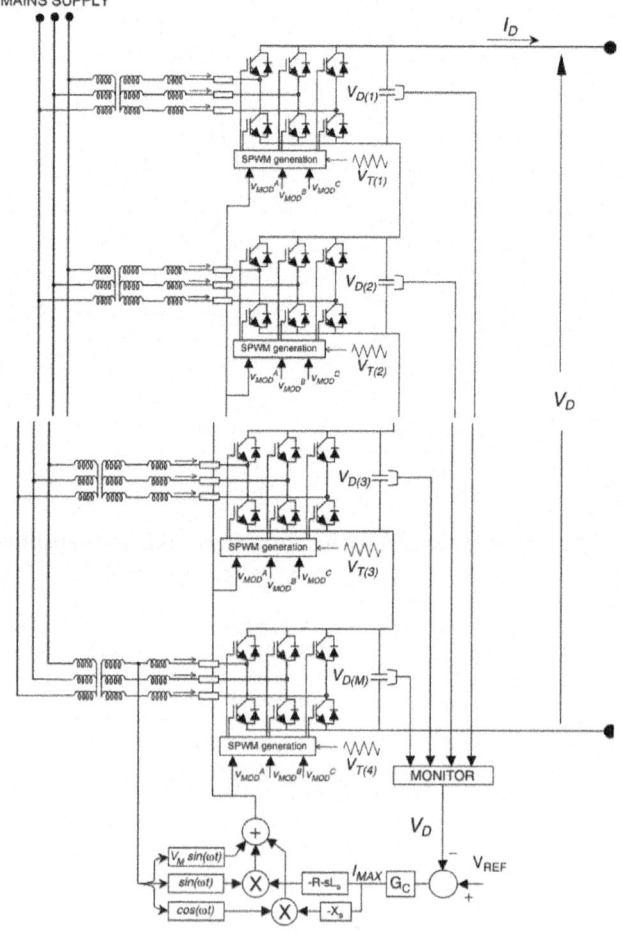

Figure 58: Series connection system with force-commutated rectifiers.

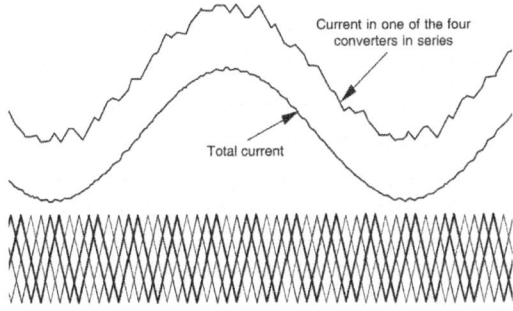

Figure 59: Input currents and carriers of the series connection system of Fig. 58.

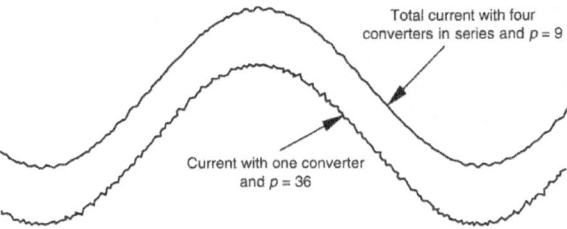

Figure 60: Four converters in series and p = 9 compared with one converter and p = 36.

where n is the number of converters in series or in parallel. It can be observed that despite the low value of p, the total current becomes quite clean and clear, better than the current of one of the converters in the chain.

The harmonic cancellation with series- or parallelconnected rectifiers, using the same modulation but the carriers shifted, is quite effective. The resultant current is better with n converters and frequency modulation $p = p_1$ than with one converter and $p = n \cdot p_1$. This attribute is verified in Fig. 60, where the total current of four converters in series with p = 9 and carriers shifted, is compared with the current of only one converter and p = 36. This technique also allows for the use of valves with slow commutation times, such as high power GTOs. Generally, high power valves have low commutation times and hence the parallel and/or series options remain very attractive.

Another speci al topology for high power was implemented for Asea Brown Boveri (ABB) in Bremen. A 100 MW power converter supplies energy to the railways at $16^{2/3}$ Hz. It uses basic "H" bridges like the one shown in Fig. 61, connected to the load through power transformers. These transformers are connected in parallel at the converter side, and in series at the load side.

The system uses SPWM with TCs shifted, and depending on the number of converters connected in the chain of bridges, the voltage waveform becomes more and more sinusoidal. Figure 62 shows a back-to-back system using a chain of 12 "H" converters connected as showed in Fig. 61b.

The ac voltage waveform obtained with the topology of Fig. 62 is displayed in Fig. 63. It can be observed that the voltage is formed by small steps that depend on the number of converters in the chain. The current is almost perfectly sinusoidal.

Figure 64 shows the voltage waveforms for different number of converters connected in the bridge. It is clear that the larger the number of converters, the better the voltage.

Another interesting result with this converter is that the ac voltages become modulated by both PWM and amplitude modulation (AM).

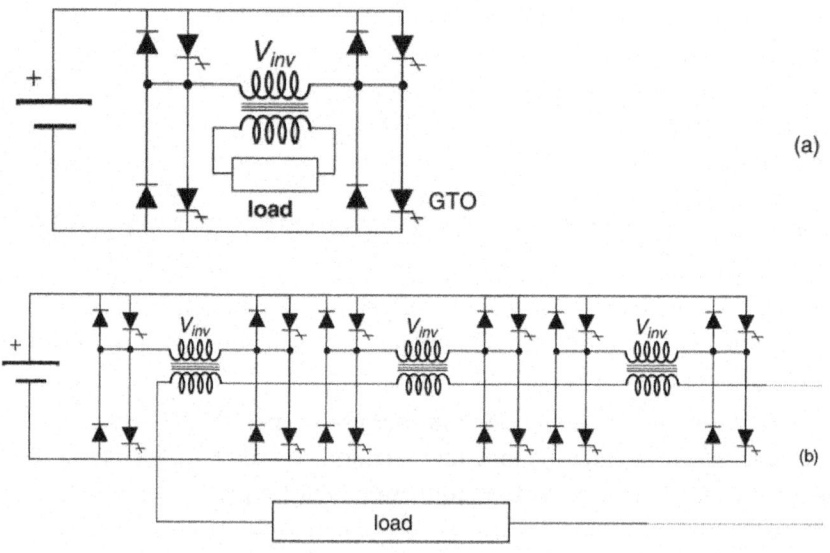

Figure 61: The "H" modulator: (a) one bridge and (b) bridges connected in series at load side through isolation transformers.

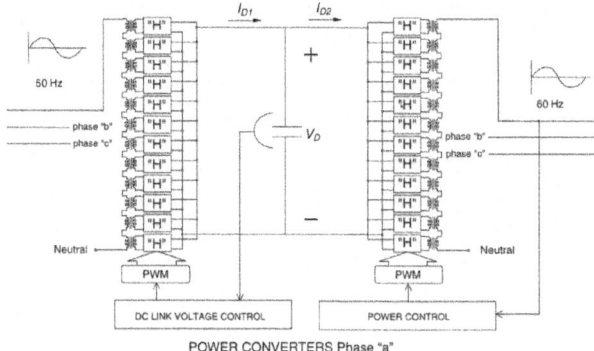

Figure 62: Frequency link with force-commutated converters and sinusoidal voltage modulation.

This is because when the pulse modulation changes, the steps of the amplitude change. The maximum number of steps of the resultant voltage is equal to the number of converters. When the voltage decreases, some steps disappear, and then the AM becomes a discrete function. Figure 65 shows the AM of the voltage.

Machine Drives Applications

One of the most important applications of force-commutated rectifiers is in machine drives. Line-commutated thyristor converters have limited applications because they need excitation to extinguish the valves. This limitation do not allow the use of line-commutated converters in induction machine drives.

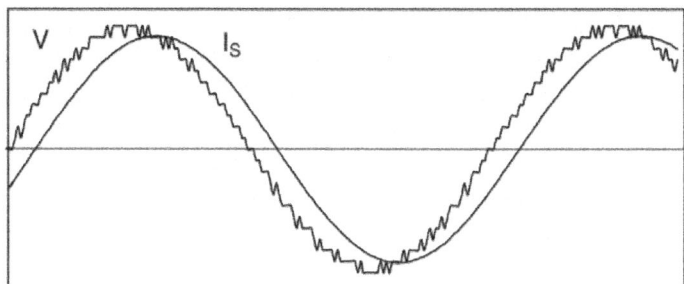

Figure 63: Voltage and current waveforms with 12 converters.

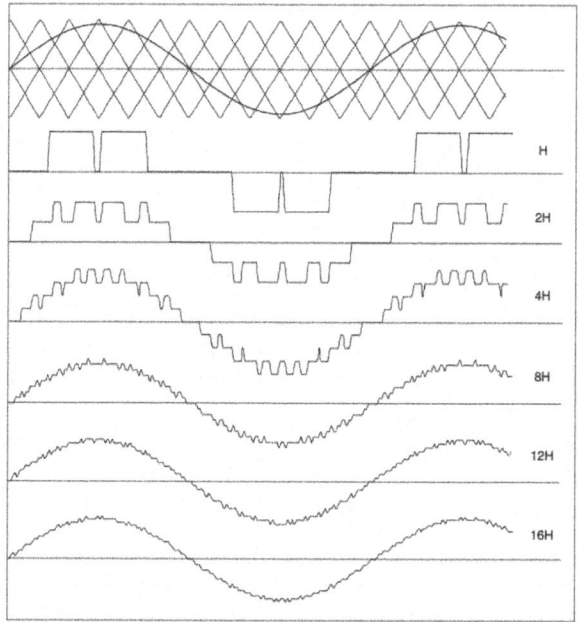

Figure 64: Voltage waveforms with different numbers of "H" bridges in series.

On the other hand, with force-commutated converters fourquadrant operation is achievable. Figure 66 shows a typical frequency converter with a force-commutated rectifier– inverter link. The rectifier side controls the dc link, and the inverter side controls the machine. The machine can be a synchronous, brushless dc, or induction machine. The reversal of both speed and power are possible with this topology. At the rectifier side, the power factor can be controlled, and even with an inductive load such as an induction machine, the source can "see" the load as capacitive or resistive. Changing the frequency of the inverter controls the machine speed, and the torque is controlled through the stator currents and torque angle.

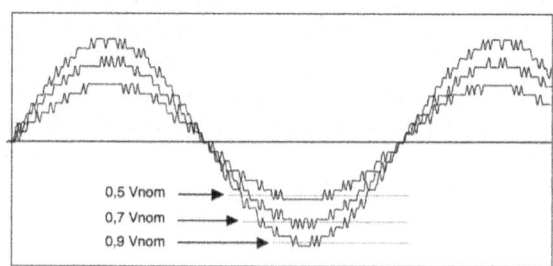

Figure 65: Amplitude modulation of the "H" bridges of Fig. 62.

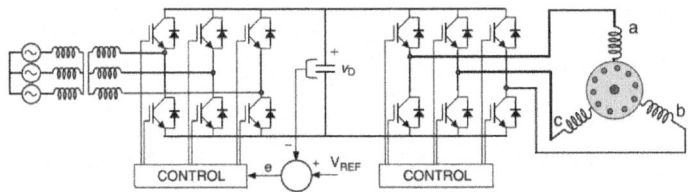

Figure 66: Frequency converter with force-commutated converters.

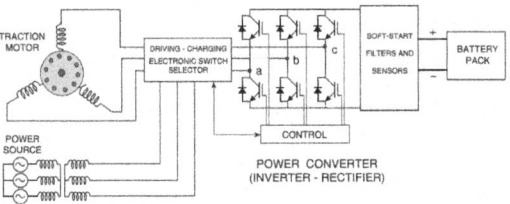

Figure 67: Electric bus system with regenerative braking and battery charger.

The inverter will become a rectifier during regenerative braking, which is possible by making slip negative in an induction machine, or by making the torque angle negative in synchronous and brushless dc machines.

A variation of the drive of Fig. 66 is found in electric traction applications. Battery powered vehicles use the inverter as a rectifier during regenerative braking, and sometimes the inverter is also used as a battery charger. In this case, the rectifier can be fed by a single-phase or three-phase system.

Figure 67 shows a battery-powered electric bus system. This system uses the power inverter of the traction motor as rectifier for two purposes: regenerative braking and as a battery charger fed by a three-phase power source.

Variable Speed Power Generation

Power generation at 50 or 60 Hz requires constant speed machines. In addition, induction machines are not currently used in power plants because of magnetization problems.

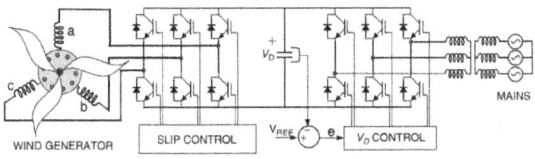

Figure 68: Variable-speed constant-frequency wind generator.

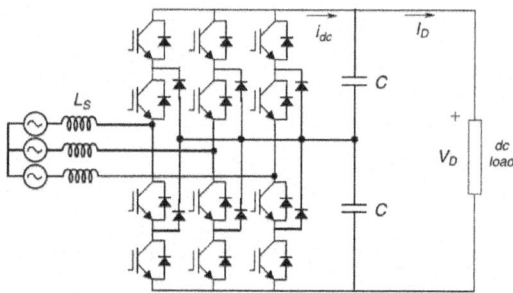

Figure 69: Voltage-source rectifier using three-level converter.

With the use of frequency-link force-commutated converters, variable-speed constant-frequency generation becomes possible even with induction generators. The power plant in Fig. 68 shows a wind generator implemented with an induction machine, and a rectifier–inverter frequency link connected to the utility. The dc link voltage is kept constant with the converter located at the mains side. The converter connected at the machine side controls the slip of the generator and adjusts it according to the speed of wind or power requirements. The utility is not affected by the power factor of the generator, because the two converters keep the cos φ of the machine independent of the mains supply. The converter at the mains side can even be adjusted to operate at leading power factor.

Varible-speed constant-frequency generation also can be used in either hydraulic or thermal plants. This allows for optimal adjustment of the efficiency-speed characteristics of the machines. In many places, wound rotor induction generators working as variable speed synchronous machines are being used as constant frequency generators. They operate in hydraulic plants that are able to store water during low demand periods. A power converter is connected at the slip rings of the generator. The rotor is then fed with variable frequency excitation. This allows the generator to generate at different speeds around the synchronous rotating flux.

Power Rectifiers Using Multilevel Topologies

Almost all voltage source rectifiers already described are twolevel configurations. Today, multilevel topologies are becoming very popular, mainly three-level converters. The most popular three-level configuration is called diode clamped converter, which is shown in Fig. 69. This topology is today the standard solution for high power steel rolling mills, which uses back-to-back three-phase rectifier–inverter link configuration. In addition, this solution has been recently introduced in high power downhill conveyor belts which operate almost permanently in the regeneration mode or rectifier operation. The more important advantage of three-level rectifiers is that voltage and current harmonics are reduced due to the increased number of levels.

Higher number of levels can be obtained using the same diode clamped strategy, as shown in Fig. 70, where only one phase of a general approach is displayed. However, this topology becomes more and more complex with the increase of number of levels. For this reason, new topologies are being studied to get a large number of levels with less power transistors.

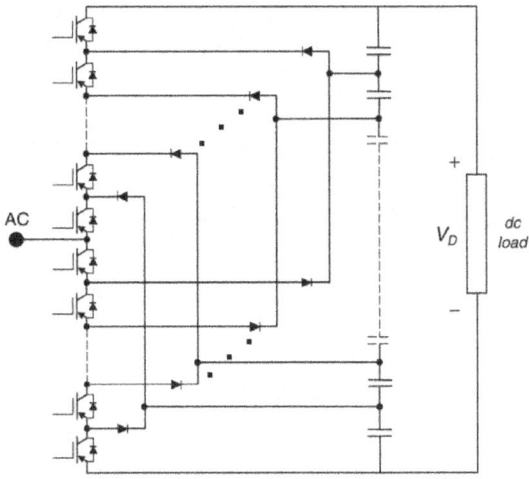

Figure 70: Multilevel rectifier using diode clamped topology.

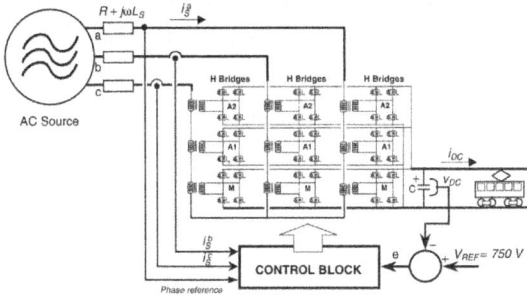

Figure 71: 27-Level rectifier for railways, using H-bridges scaled in power of three.

One example of such of these topologies is the multistage, 27-level converter shown in Fig. 71. This special 27-level, four-quadrant rectifier, uses only three H-bridges per phase with independent input transformers for each H-bridge. The transformers allow galvanic isolation and power escalation to get high quality voltage waveforms, with THD of less than 1%. The power scalation consists on increasing the voltage rates of each transformer making use of the "three-level" characteristics of H-bridges. Then, the number of levels is optimized when transformers are scaled in power of three. Some advantages of this 27-level topology are: (a) only one of the three H-bridges, called "main converter," manages more than 80% of the total active power in each phase and (b) this main converter switches at fundamental frequency reducing the switching losses at a minimum value. The rectifier of Fig. 71 is a current-controlled voltage source type, with a conventional feedback control loop, which is being used as a rectifier in a subway substation. It includes fast reversal of power and the ability to produce clean ac and dc waveforms with negligible ripple. This rectifier can also compensate power factor and eliminate harmonics produced by other loads in the ac line. Figure 72 shows the ac voltage waveform obtained with this rectifier from an experimental prototype. If one more H-bridge is added, 81 levels are obtained, because the number of levels increases according with $N = 3^k$, where N is the number of levels or voltage steps and k the number of H-bridges used per phase.

Many other high-level topologies are under study but this matter is beyond the main topic of this chapter.

REFERENCES

1. G. Möltgen, "Line Commutated Thyristor Converters," Siemens Aktiengesellschaft, Berlin-Munich, Pitman Publishing, London, 1972.

2. G. Möltgen, "Converter Engineering, and Introduction to Operation and Theory," John Wiley and Sons, New York, 1984.

3. K. Thorborg, "Power Electronics," Prentice-Hall International (UK) Ltd., London, 1988.

4. M. H. Rashid, "Power Electronics, Circuits Devices and Applications," Prentice-Hall International Editions, London, 1992.

5. N. Mohan, T. M. Undeland, and W. P. Robbins, "Power Electronics: Converters, Applications, and Design," John Wiley and Sons, New York 1989.

6. J. Arrillaga, D. A. Bradley, and P. S. Bodger, "Power System Harmonics," John Wiley and Sons, New York, 1989.

7. J. M. D. Murphy and F. G. Turnbull, "Power Electronic Control of AC Motors," Pergamon Press, 1988.

8. M. E. Villablanca and J. Arrillaga, "Pulse Multiplication in Parallel Convertors by Multitap Control of Interphase Reactor," IEE Proceedings-B, Vol. 139, No 1; January 1992, pp. 13–20.

9. D. A. Woodford, "HVDC Transmission," Professional Report from Manitoba HVDC Research Center, Winnipeg, Manitoba, March 1998.

10. D. R. Veas, J. W. Dixon, and B. T. Ooi, "A Novel Load Current Control Method for a Leading Power Factor Voltage Source PEM Rectifier," IEEE Transactions on Power Electronics, Vol. 9, No 2, March 1994, pp. 153–159.

11. L. Morán, E. Mora, R. Wallace, and J. Dixon, "Performance Analysis of a Power Factor Compensator which Simultaneously Eliminates Line Current Harmonics," IEEE Power Electronics Specialists Conference, PESC'92, Toledo, España, June 29 to July 3, 1992.

12. P. D. Ziogas, L. Morán, G. Joos, and D. Vincenti, "A Refined PWM Scheme for Voltage and Current Source Converters," IEEE-IAS Annual Meeting, 1990, pp. 977–983.

13. W. McMurray, "Modulation of the Chopping Frequency in DC Choppers and PWM Inverters Having Current Hysteresis Controllers," IEEE Transaction on Ind. Appl., Vol. IA-20, July/August 1984, pp. 763–768.

14. J. W. Dixon and B. T. Ooi, "Indirect Current Control of a Unity Power Factor Sinusoidal Current Boost Type Three-Phase Rectifier," IEEE Transactions on Industrial Electronics, Vol. 35, No 4, November 1988, pp. 508–515.

15. L. Morán, J. Dixon, and R. Wallace "A Three-Phase Active Power Filter Operating with Fixed Switching Frequency for Reactive Power and Current Harmonic Compensation," IEEE Transactions on Industrial Electronics, Vol. 42, No 4, August 1995, pp. 402–408.

16. M. A. Boost and P. Ziogas, "State-of-the-Art PWM Techniques, a Critical Evaluation," IEEE Transactions on Industry Applications, Vol. 24, No 2, March/April 1988, pp. 271–280.

17. J. W. Dixon and B. T. Ooi, "Series and Parallel Operation of Hysteresis Current-Controlled PWM Rectifiers," IEEE Transactions on Industry Applications, Vol. 25, No 4, July/August 1989, pp. 644–651.

18. B. T. Ooi, J. W. Dixon, A. B. Kulkarni, and M. Nishimoto, "An integrated AC Drive System Using a Controlled-Current PWM Rectifier/Inverter Link," IEEE Transactions on Power Electronics, Vol. 3, No 1, January 1988, pp. 64–71.

19. M. Koyama, Y. Shimomura, H. Yamaguchi, M. Mukunoki, H. Okayama, and S. Mizoguchi, "Large Capacity High Efficiency Three-Level GCT Inverter System for Steel Rolling Mill Drivers," Proceedings of the 9th European Conference on Power Electronics, EPE 2001, Austria, CDROM.

20. J. Rodríguez, J. Dixon, J. Espinoza, and P. Lezana, "PWM Regenerative Rectifiers: State of the Art," IEEE Transactions on Industrial Electronics, Vol. 52, No 4, January/February 2005, pp. 5–22.

21. J. Dixon and L. Morán, "A Clean Four-Quadrant Sinusoidal Power Rectifier, Using Multistage Converters for Subway Applications," IEEE Transactions on Industrial Electronics, Vol. 52, No 5, May–June 2005, pp. 653–661.

Chapter 7

RESONANT AND SOFT-SWITCHING CONVERTERS

INTRODUCTION

Advances in power electronics in the last few decades have led to not just improvements in power devices, but also new concepts in converter topologies and control. In the 1970s, conventional pulse width modulated (PWM) power converters were operated in a switched mode operation. Power switches have to cut off the load current within the turn-on and turn-off times under the hard switching conditions. Hard switching refers to the stressful switching behavior of the power electronic devices. The switching trajectory of a hardswitched power device is shown in Fig. 1. During the turn-on and turn-off processes, the power device has to withstand high voltage and current simultaneously, resulting inhigh switching losses and stress. Dissipative passive snubbers are usually added to the power circuits so that the dv/dt and di/dt of the power devices could be reduced, and the switching loss and stress be diverted to the passive snubber circuits. However, the switching loss is proportional to the switching frequency, thus limiting the maximum switching frequency of the power converters. Typical converter switching frequency was limited to a few tens of kilo-Hertz (typically 20–50 kHz) in early 1980s. The stray inductive and capacitive components in the power circuits and power devices still cause considerable transient effects, which in turn give rise to electromagnetic interference (EMI) problems.

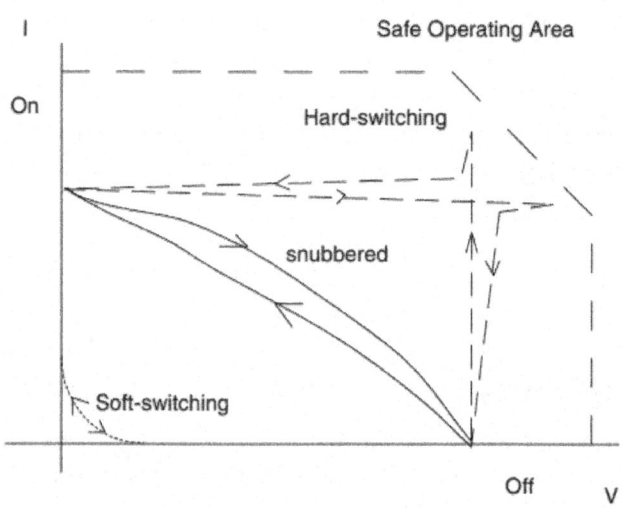

Figure 1: Typical switching trajectories of power switches.

Figure 2 shows idealswitching waveforms and typical practical waveforms of the switch voltage. The transient ringing effects are major causes of EMI.

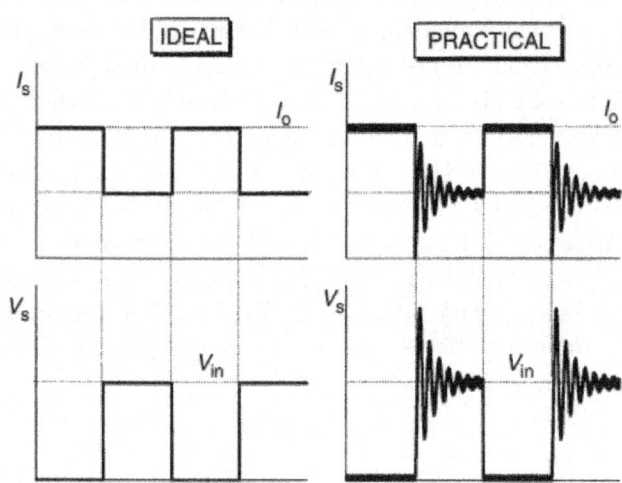

Figure 2: Typical: (a) ideal and (b) practical switching waveforms.

In the 1980s, lots of research efforts were diverted towards the use of resonant converters. The concept was to incorporate resonant tanks in the converters to create oscillatory (usually sinusoidal) voltage and/or current waveforms so that zero-voltage switching (ZVS) or zero-current switching (ZCS) conditions can be created for the power switches. The reduction of switching loss and the continual improvement of power switches allow the switching frequency of the resonant converters to reach hundreds of kilo-Hertz (typically 100–500 kHz). Consequently, the size of magnetic components can be reduced and the power density of the converters increased. Various forms of resonant converters have been proposed and developed. However, most of the resonant converters suffer several problems. When compared with the conventional PWM converters, the resonant current and the voltage of resonant converters have high peak values, leading to higher conduction loss and higher V and I rating requirements for the power devices. Also, many resonant converters require frequency modulation (FM) for output regulation. Variable switching frequency operation makes the filter design and control more complicated.

In late 1980s and throughout 1990s, further improvements have been made in converter technology. New generations of soft-switched converters that combine the advantages of conventional PWM converters and resonant converters have been developed. These soft-switched converters have switching waveforms similar to those of conventional PWM converters except that the rising and falling edges of the waveforms are "smoothed" with no transient spikes. Unlike the resonant converters, new soft-switched converters usually utilize the resonance in a controlled manner. Resonance is allowed to occur just before and during the turn-on and turn-off processes so as to create ZVS and ZCS conditions. Other than that, they behave just like conventional PWM converters. With simple modifications, many customized control integrated circuits (ICs) designed for conventional converters can be employed for soft-switched converters. Because the switching loss and stress have been reduced, soft-switched converter can be operated at the very high frequency (typically 500 kHz to a few Mega-Hertz). Soft-switching converters also provide an effective solution to suppress EMI and have been applied to DC–DC, AC–DC, and DC–AC converters. This chapter covers the basic technology of resonant and soft-switching converters. Various forms of soft-switching techniques such as ZVS, ZCS, voltage clamping, zero-voltage transition methods, etc. are addressed. The emphasis is placed on the basic operating principle and practicality of the converters without using much mathematical analysis.

CLASSIFICATION

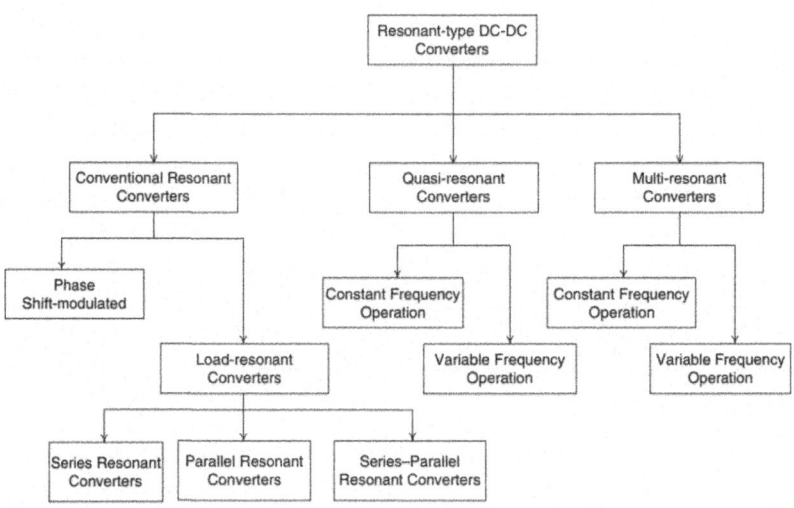

RESONANT SWITCH

Prior to the availability of fully controllable power switches, thyristors were the major power devices used in power electronic circuits. Each thyristor requires a commutation circuit, which usually consists of a LC resonant circuit, for forcing the current to zero in the turn-off process [1]. This mechanism is in fact a type of zero-current turn-off process. With the recent advancement in semiconductor technology, the voltage and current handling capability, and the switching speed of fully controllable switches have significantly been improved. In many high power applications, controllable switches such as gate turn-offs (GTOs) and insulated gate bipolar transistors (IGBTs) have replaced thyristors [2, 3]. However, the use of resonant circuit for achieving ZCS and/ or ZVS [4–8] has also emerged as a new technology for power converters. The concept of resonant switch that replaces conventional power switch is introduced in this section.

A resonant switch is a sub-circuit comprising a semiconductor switch S and resonant elements, L_r and C_r [9–11]. The switch S can be implemented by a unidirectional or bidirectional switch, which determines the operation mode of the resonant switch. Two types of resonant switches [12], including zero-current (ZC) resonant switch and zero-voltage (ZV) resonant switches, are shown in Figs. 3 and 4, respectively.

ZC Resonant Switch

In a ZC resonant switch, an inductor L_r is connected in series with a power switch S in order to achieve zero-current switching (ZCS). If the switch S is a unidirectional switch, the switch current is allowed to resonate in the positive half cycle only. The resonant switch is said to operate in half-wave mode. If a diode is connected in anti-parallel with the unidirectional switch, the switch current can flow in both directions. In this case, the resonant switch can operate in full-wave mode. At turn-on, the switch current will rise slowly from zero. It will then oscillate, because of the resonance between L_r and C_r. Finally, the switch can be commutated at the next zero current duration. The objective of this type of switch is to shape the switch current waveform during conduction time in order to create a zero-current condition for the switch to turn off [13].

ZV Resonant Switch

In a ZV resonant switch, a capacitor Cr is connected in parallel with the switch S for achieving zero-voltage switching (ZVS).

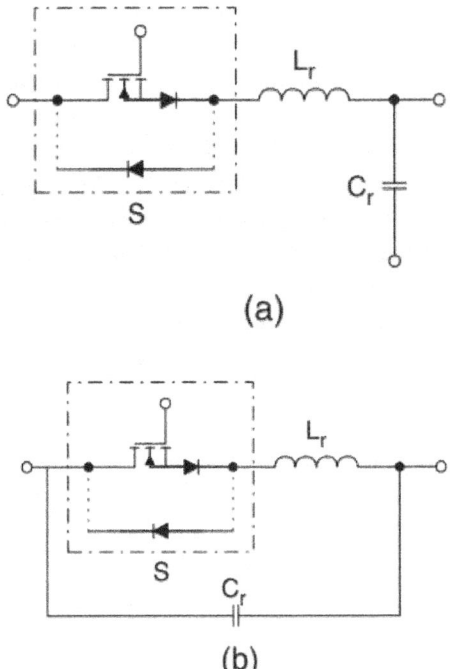

(a)

(b)

Figure 3: Zero-current (ZC) resonant switch.

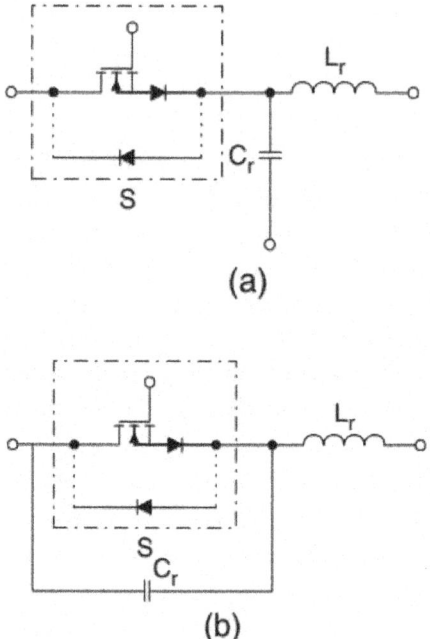

Figure 4: Zero-voltage (ZV) resonant switch.

If the switch S is a unidirectional switch, the voltage across the capacitor C_r can oscillate freely in both positive and negative half-cycle. Thus, the resonant switch can operate in full-wave mode. If a diode is connected in anti-parallel with the unidirectional switch, the resonant capacitor voltage is clamped by the diode to zero during the negative half-cycle. The resonant switch will then operate in half-wave mode. The objective of a ZV switch is to use the resonant circuit to shape the switch voltage waveform during the off time in order to create a zero-voltage condition for the switch to turn on [13].

QUASI-RESONANT CONVERTERS

Quasi-resonant converters (QRCs) can be considered as a hybrid of resonant and PWM converters. The underlying principle is to replace the power switch in PWM converters with the resonant switch. A large family of conventional converter circuits can be transformed into their resonant converter counterparts. The switch current and/or voltage waveforms are forced to oscillate in a quasi-sinusoidal manner, so that ZCS and/or ZVS can be achieved. Both ZCS-QRCs and ZVS-QRCs have half-wave and full-wave mode of operations [8–10, 12].

ZCS-QRCs

A ZCS-QRC designed for half-wave operation is illustrated with a buck type DC–DC converter. The schematic is shown in Fig. 5a. It is formed by replacing the power switch in conventional PWM buck converter with the ZC resonant switch in Fig. 3a. The circuit waveforms in steady state are shown in Fig. 5b. The output filter inductor L_f is sufficiently large sothat its current is approximately constant. Prior to turning the switch on, the output current Io freewheels through the output diode D_f . The resonant capacitor voltage V_{Cr} equals zero. At t_0, the switch is turned on with ZCS. A quasi-sinusoidal current I_S flows through L_r and C_r, the output filter, and the load. S is then softly commutated at t_1 with ZCS again. During and after the gate pulse, the resonant capacitor voltage V_{Cr} rises and then decays at a rate depending on the output current. Output voltage regulation is achieved by controlling the switching frequency. Operation and characteristics of the converter depend mainly on the design of the resonant circuit L_rC_r. The following parameters are defined: voltage conversion ratio M, characteristic impedance Z_r, resonant frequency f_r, normalized load resistance r, normalized switching frequency γ.

$$M = \frac{V_o}{V_i} \tag{1a}$$

$$Z_r = \sqrt{\frac{L_r}{C_r}} \tag{1b}$$

$$f_r = \frac{1}{2\pi\sqrt{L_r C_r}} \tag{1c}$$

$$r = \frac{R_L}{Z_r} \tag{1d}$$

$$\gamma = \frac{f_s}{f_r} \tag{1e}$$

It can be seen from the waveforms that if $I_o > V_i/Z_r$, I_S will not come back to zero naturally and the switch will have to beforced off, thus resulting in turn-off losses. The relationships between M and γ at different r are shown in Fig. 5c. It can be seen that M is sensitive to the load variation. At light load conditions, the unused energy is stored in C_r, leading to an increase in the output voltage. Thus, the switching frequency has to be controlled, in order to regulate the output voltage.

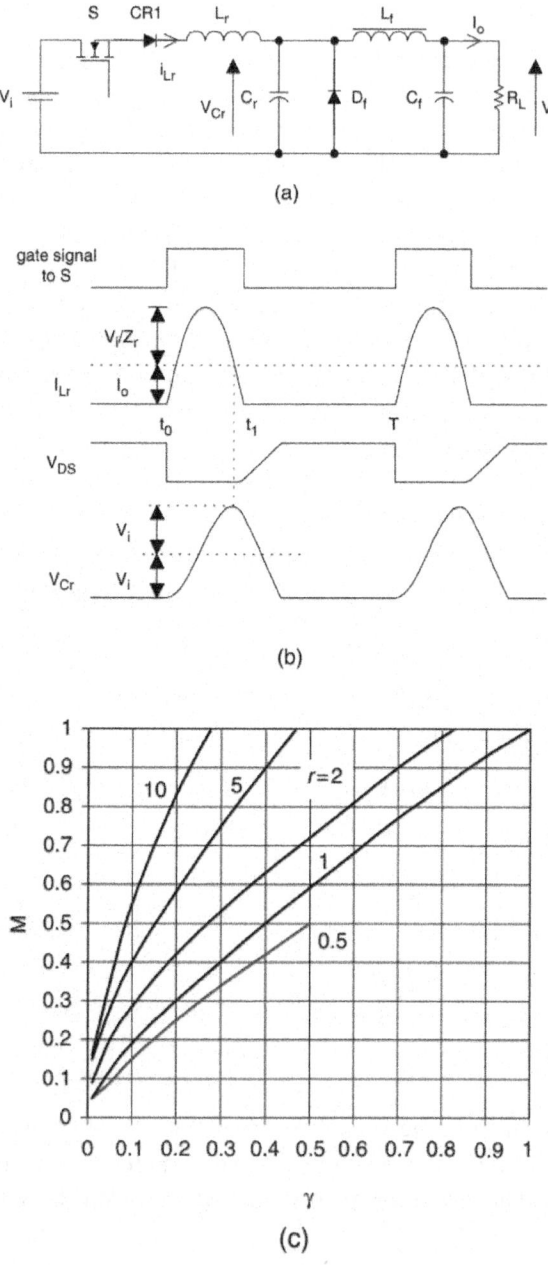

(a)

(b)

(c)

Figure 5: Half-wave, quasi-resonant buck converter with ZCS: (a) schematic diagram; (b) circuit waveforms; and (c) relationship between M and γ.

If an anti-parallel diode is connected across the switch, the converter will be operating in full-wave mode. The circuit schematic is shown in Fig. 6a. The circuit waveforms in steady state are shown in Fig. 6b. The operation is similar to the one in half-wave mode. However, the inductor current is allowed to reverse through the anti-parallel diode and the duration for the resonant stage is lengthened. This permits excess energy in the resonant circuit at light loads to be transferred back to the voltage source V_i. This significantly reduces the dependence of V_o on the output load. The relationships between M and γ at different r are shown in Fig. 6c. It can be seen that M is insensitive to load variation.

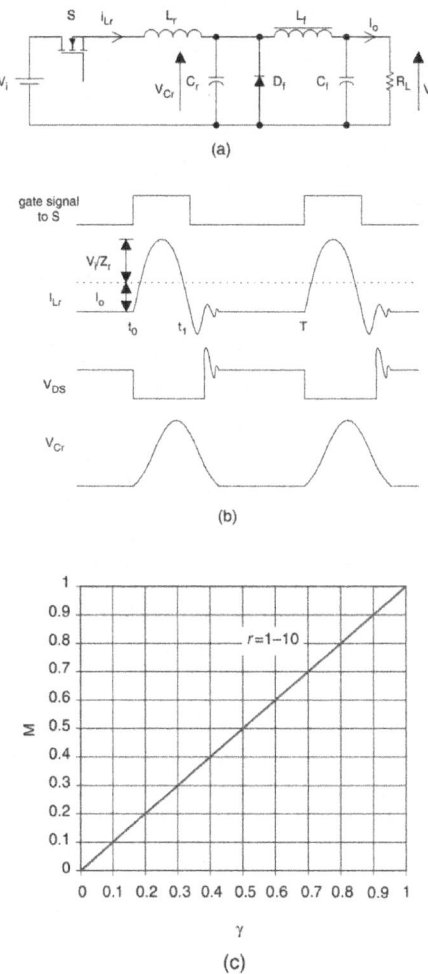

Figure 6: Full-wave, quasi-resonant buck converter with ZCS: (a) schematic diagram; (b) circuit waveforms; and (c) relationship between M and γ.

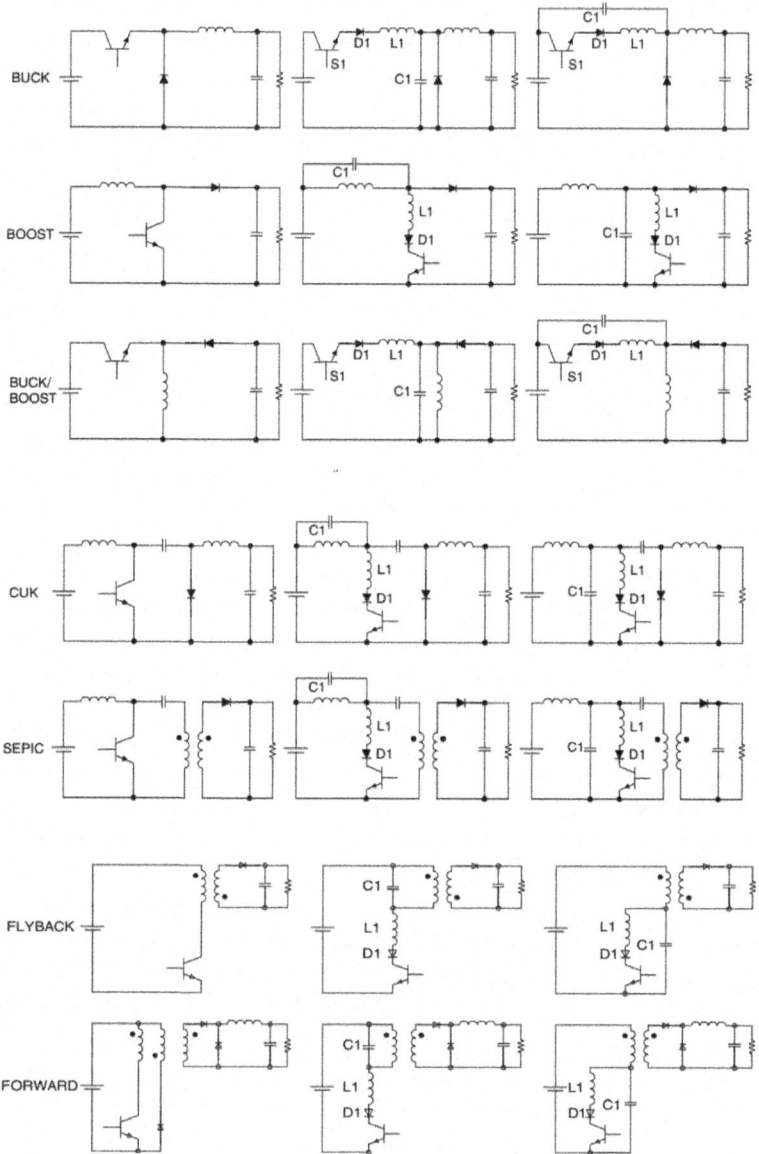

Figure 7: A family of quasi-resonant converter with ZCS.

A quasi-resonant buck converter designed for half-wave operation is shown in Fig. 8a – using a ZV resonant switch in Fig. 4b. The steady-state circuit waveforms are shown in Fig. 8b. Basic relations of ZVS-QRCs are given in Eqs. (1a–e). When the switch S is turned on, it carries the output current Io. The supply voltage V_i reverse biases the diode D_f. When the switch is zero-voltage (ZV) turned off, the output current starts to flow through the resonant capacitor C_r. When the resonant capacitor voltage V_{Cr} is equal to V_i, D_f turns on. This starts the resonant stage. When V_{Cr} equals zero, the anti-parallel diode turns on. The resonant capacitor is shorted and the source voltage is applied to the resonant inductor L_r. The resonant inductor current I_{Lr} increases linearly until it reaches I_o. Then D_f turns off. In order to achieve ZVS, S should be triggered during the time when the antiparallel diode conducts. It can be seen from the waveforms that the peak amplitude of the resonant capacitor voltage should be greater or equal to the input voltage (i.e. $I_o Z_r > V_{in}$). From Fig. 8c, it can be seen that the voltage conversion ratio is load-sensitive. In order to regulate the output voltage for different loads r, the switching frequency should also be changed accordingly.

ZVS converters can be operated in full-wave mode. The circuit schematic is shown in Fig. 9a. The circuit waveforms in steady state are shown in Fig. 9b. The operation is similar to half-wave mode of operation, except that V_{Cr} can swing between positive and negative voltages. The relationships between M and g at different r are shown in Fig. 9c.

Comparing Fig. 8c with Fig. 9c, it can be seen that M is load-insensitive in full-wave mode. This is a desirable feature. However, as the series diode limits the direction of the switch current, energy will be stored in the output capacitance of the switch and will dissipate in the switch during turn on. Hence, the full-wave mode has the problem of capacitive turn-on loss, and is less practical in high frequency operation. In practice, ZVS-QRCs are usually operated in half-wave mode rather than full-wave mode.

By replacing the ZV resonant switch in the conventional converters, various ZVS-QRCs can be derived. They are shown in Fig. 10.

Comparisons between ZCS and ZVS

ZCS can eliminate the switching losses at turn off and reduce the switching losses at turn on. As a relatively large capacitor is connected across the output diode during resonance, the converter operation becomes insensitive to the diode's junction capacitance. When power MOSFETs are zero-current switched on, the energy stored in the device's capacitance will be dissipated. This capacitive turn-on loss is proportional to the switching frequency. During turn on, considerable rate of change of voltage can be coupled to the gate drive circuit through the Miller capacitor, thus increasing switching loss and noise. Another limitation is that the switches are under highcurrent stress, resulting in higher conduction loss. However, it should be noted that ZCS is particularly effective in reducing switching loss for power devices (such as IGBT) with large tail current in the turn-off process.

ZVS eliminates the capacitive turn-on loss. It is suitable for high-frequency operation. For single-ended configuration, the switches could suffer from excessive voltage stress, which is proportional to the load. It will be shown in Section 5 that the maximum voltage across switches in half-bridge and fullbridge configurations is clamped to the input voltage.

For both ZCS and ZVS, output regulation of the resonant converters can be achieved by variable frequency control. ZCS operates with constant on-time control, while ZVS operates with constant off-time control. With a wide input and load range, both techniques have to operate with a wide switching frequency range, making it not easy to design resonant converters optimally.

ZVS IN HIGH FREQUENCY APPLICATIONS

By the nature of the resonant tank and ZCS, the peak switch current in resonant converters is much higher than that in the square-wave counterparts. In addition, a high voltage will be established across the switch in the off state after the resonant stage. When the switch is switched on again, the energy stored in the output capacitor will be discharged through the switch, causing a significant power loss at high frequencies and high voltages. This switching loss can be reduced by using ZVS.

ZVS can be viewed as square-wave power utilizing a constant off-time control. Output regulation is achieved by controlling the on time or switching frequency. During the off time, the resonant tank circuit traverses the voltage across the switch from zero to its peak value and then back to zero again. At that ZV instant, the switch can be reactivated. Apart from the conventional single-ended converters, some other examples of converters with ZVS are illustrated in the following section.

ZVS with Clamped Voltage

The high voltage stress problem in the single-switch configuration with ZVS can be avoided in half-bridge (HB) and full-bridge (FB) configurations [14–17]. The peak switch voltage can be clamped to the dc supply rail, and thus reducing the switch voltage stress. In addition, the series transformer leakage and circuit inductance can form parts of the resonant path. Therefore, these parasitic components, which are undesirable in hard-switched converter become useful components in ZVS ones. Figures 11 and 12 show the ZVS HB and FB circuits, respectively, together with the circuit waveforms.

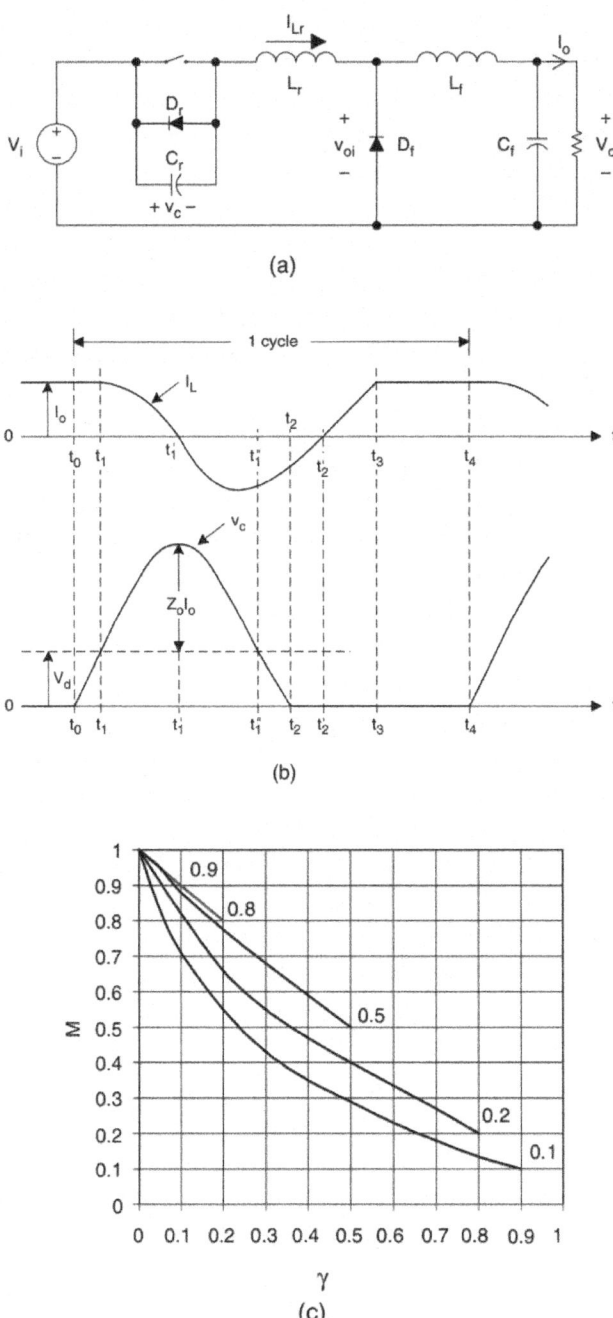

Figure 8: Half-wave, quasi-resonant buck converter with ZVS: (a) schematic diagram; (b) circuit waveforms; and (c) relationship between M and γ.

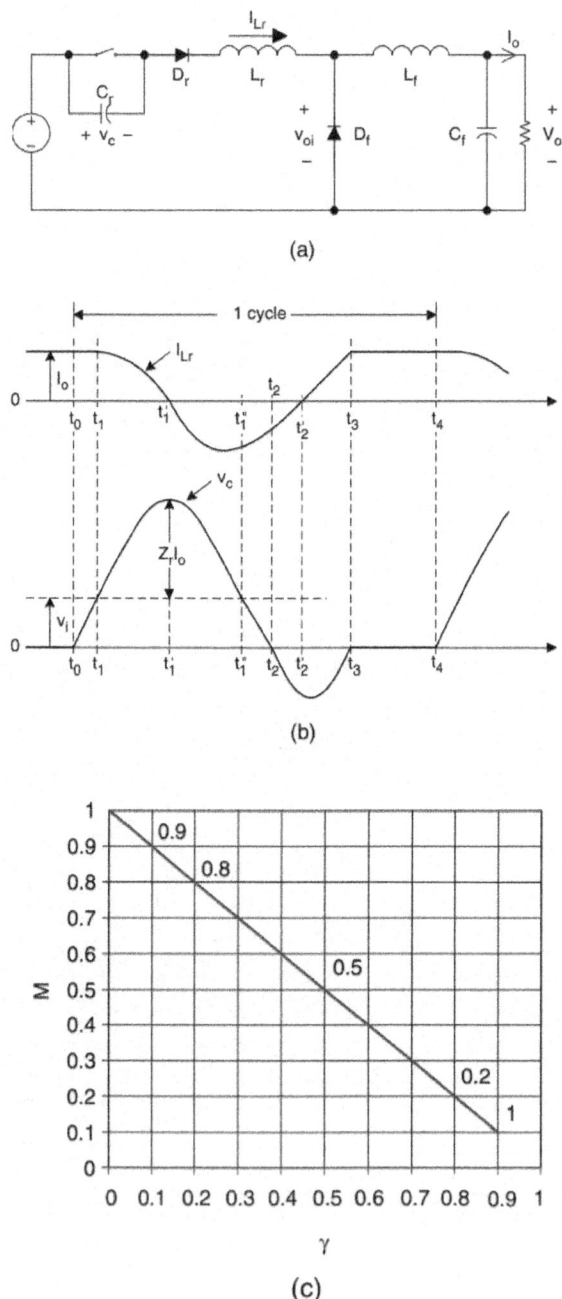

Figure 9: Full-wave, quasi-resonant buck converter with ZVS: (a) schematic diagram; (b) circuit waveforms; and (c) relationship between M and γ.

The resonant capacitor is equivalent to the parallel connection of the two capacitors ($C_r/2$) across the switches. The off-state voltage of the switches will not exceed the input voltage during resonance because they will be clamped to the supply rail by the anti-parallel diode of the switches.

Phase-shifted Converter with Zero Voltage Transition

In a conventional FB converter, the two diagonal switch pairs are driven alternatively. The output transformer is fed with anac rectangular voltage. By applying a phase-shifting approach, a deliberate delay can be introduced between the gate signals to the switches [18].

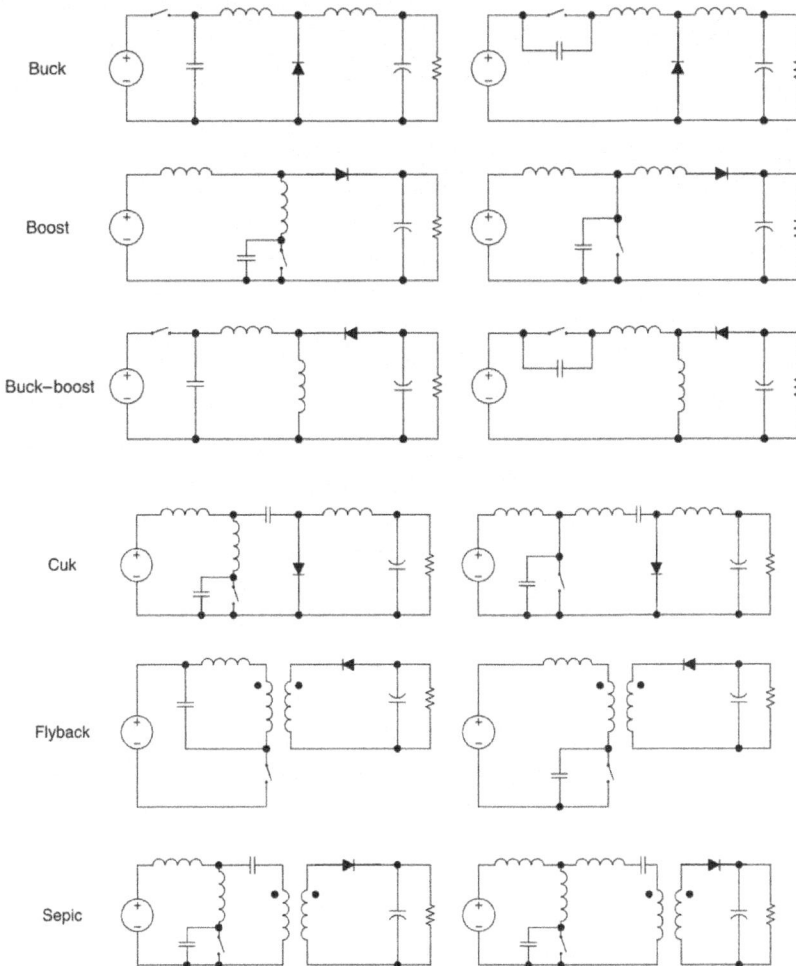

Figure 10: A family of quasi-resonant converter with ZVS.

The circuit waveforms are shown in Fig. 13. Two upper or lower switches can be conducting (either through the switch or the anti-parallel diode), yet the applied voltage to the transformer is zero. This zero-voltage condition appears in the interval $[t_1, t_2]$ of V_{pri} in Fig. 13. This operating stage corresponds to the required off time for that particular switching cycle. When the desired switch is turned off, the primary transformer current flows into the switch output capacitance causing the switch voltage to resonate to the opposite input rail. Effects of the parasitic circuit components are used advantageously to facilitate the resonant transitions. This enables a ZVS condition for turning on the opposite switch. Thus, varying the phase shift controls the effective duty cycle and hence the output power. The resonant circuit is necessary to meet the requirement of providing sufficient inductive energy to drive the capacitors to the opposite bus rail. The resonant transition must be achieved within the designed transition time.

MULTI-RESONANT CONVERTERS (MRC)

The ZCS- and ZVS-QRCs optimize the switching condition for either the active switch or the output diode only, but notfor both of them simultaneously. Multi-resonant switch concept, which is an extension of the concept of the resonant switch, has been developed to overcome such limitation. The zero-current multi-resonant (ZC-MR) and zero-voltage multiresonant (ZV-MR) switches [12, 17] are shown in Fig. 14.

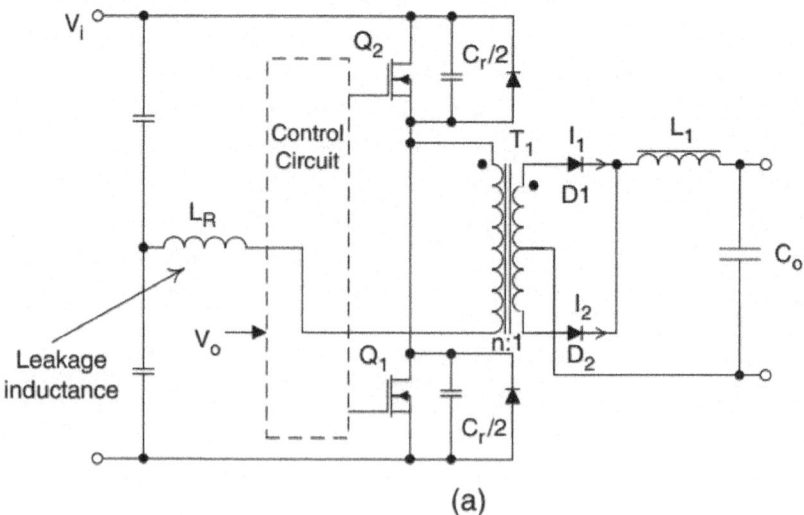

(a)

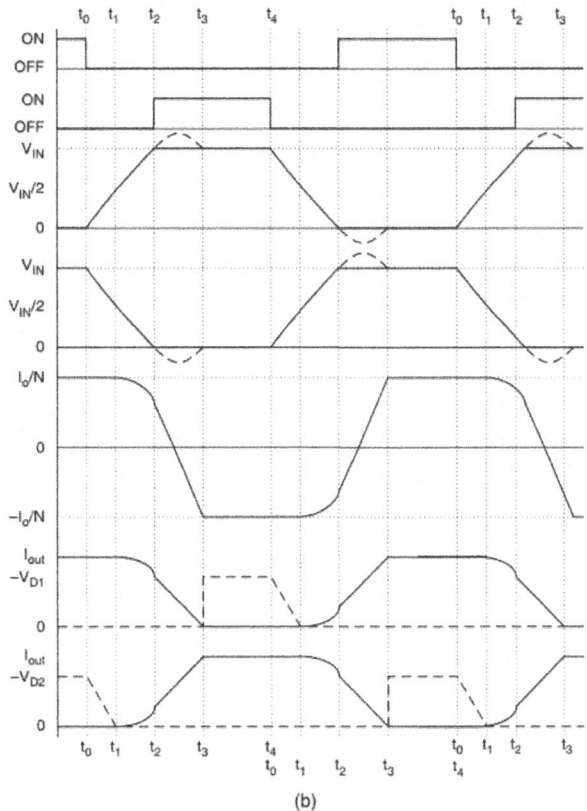

(b)

Figure 11: Half-bridge converter with ZVS: (a) circuit diagram and (b) circuit waveforms.

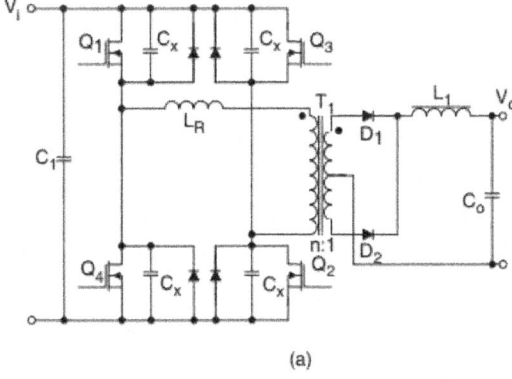

(a)

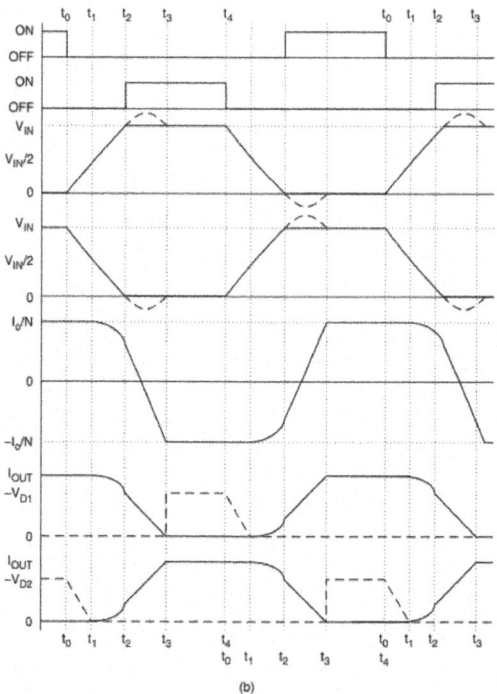

Figure 12: Full-bridge converter with ZVS: (a) circuit schematics and (b) circuit waveforms.

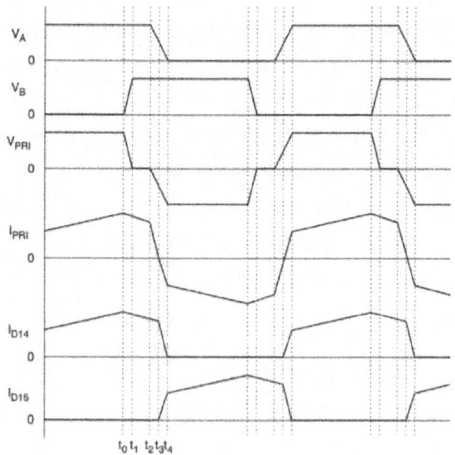

Figure 13: Circuit waveforms of the phase-shifted, ZVT FB converter.

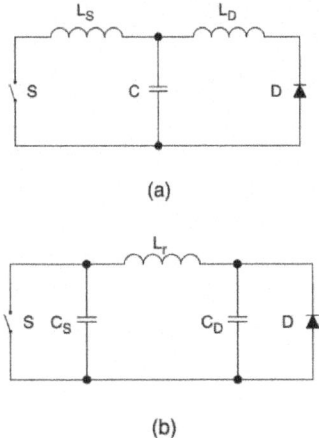

(a)

(b)

Figure 14: Multi-resonant switches: (a) ZC-MR switch and (b) ZV-MR switch.

The multi-resonant circuits incorporate all major parasitic components, including switch output capacitance, diode junction capacitance, and transformer leakage inductance into the resonant circuit. In general, ZVS (half-wave mode) is more favorable than ZCS in DC–DC converters for high-frequency operation because the parasitic capacitance of the active switch and the diode will form a part of the resonant circuit.

An example of a buck ZVS-MRC is shown in Fig. 15. Depending on the ratio of the resonant capacitance C_D/C_S, two possible topological modes, namely mode I and mode II, can be operated [19]. The ratio affects the time at which the voltages across the switch S and the output diode D_f become zero. Their waveforms are shown in Figs. 16a and b, respectively. If diode voltage V_D falls to zero earlier than the switchvoltage V_S, the converter will follow mode I. Otherwise, the converter will follow mode II.

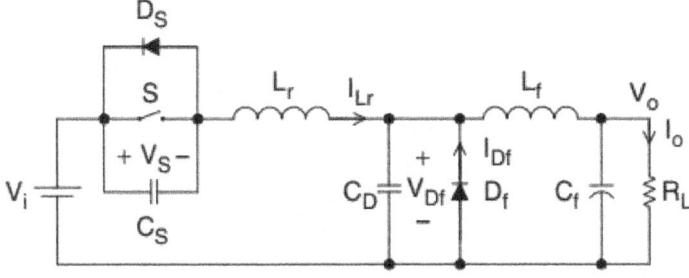

Figure 15: Buck ZVS-MRC.

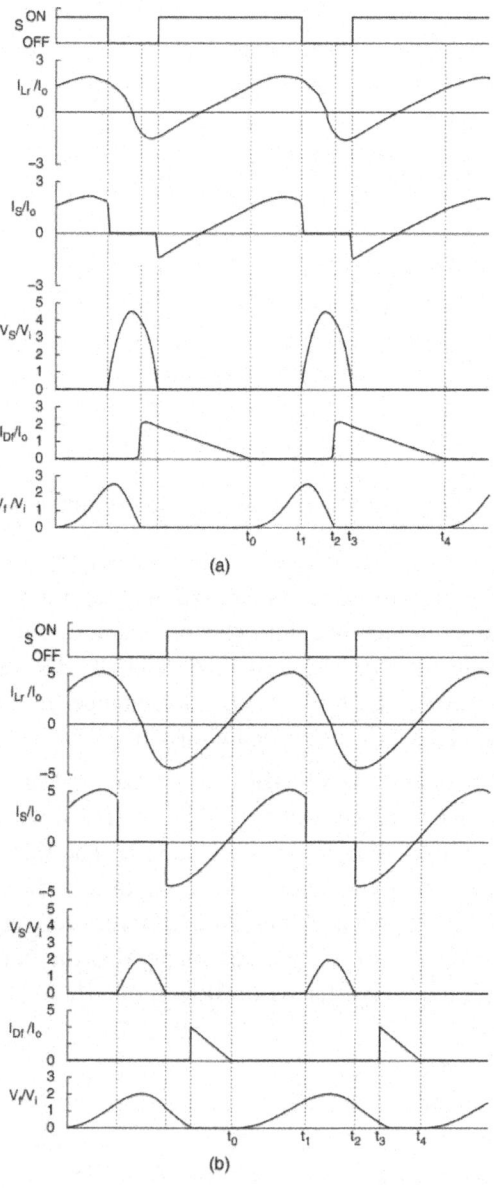

Figure 16: Possible modes of the buck ZVS-MRC: (a) mode I and (b) mode II.

Instead of having one resonant stage, there are three in this converter. The mode I operation in Fig. 16a is described first. Before the switch S is turned on, the output diode D_f is conducting and the resonant inductor current I_{Lr} is negative (flowing through the anti-parallel diode of S). S is then turned on with ZVS. The resonant inductor current I_{Lr} increases linearly and D_f is still conducting. When I_{Lr} reaches the output current I_o, the first resonant stage starts. The resonant circuit is formed by the resonant inductor Lr and the capacitor C_D across the output diode. This stage ends when S is turned off with ZVS. Then, a second resonant stage starts. The resonant circuit consists of L_r, C_D, and the capacitor across the switch Cs. This stage ends when the output diode becomes forward biased. A third resonant stage will then start. L_r and C_s form the resonant circuit. This stage ends and completes one operation cycle when the diode C_s becomes forward biased.

The only difference between mode I and mode II in Fig. 16b is in the third resonant stage, in which the resonant circuit is formed by L_r and C_D. This stage ends when D_f becomes forward biased. The concept of the multi-resonant switch can be applied to conventional converters [19–21]. A family of MRCs are shown in Fig. 17.

Although the variation of the switching frequency for regulation in MRCs is smaller than that of QRCs, a wide-band frequency modulation is still required. Hence, the optimal design of magnetic components and the EMI filters in MRCs is not easy. It would be desirable to have a constant switching frequency operation. In order to operate the MRCs with constant switching frequency, the diode in Fig. 14 can be replaced with an active switch S_2 [22]. A constant-frequency multi-resonant (CF-MR) switch is shown in Fig. 18. The output voltage is regulated by controlling the on-time of the two switches. This concept can be illustrated with the buck converter as shown in Fig. 19, together with the gate drive waveforms and operating stages. S_1 and S_2 are turned on during the time when currents flow through the antiparallel diodes of S_1 and S_2. This stage ends when S_2 is turned off with ZVS. The first resonant stage is then started. Lr and CS2 form the resonant circuit. A second resonant stage begins. L_r resonates with C_{S1} and C_{S2}. The voltage across S1 oscillates to zero. When I_{Lr} becomes negative, S1 will be turned on with ZVS. Then, Lr resonates with C_{S2}. S_2 will be turned on when current flows through D_{S2}. As the output voltage is the average voltage across S_2, output voltage regulation is achieved by controlling the conduction time of S_2.

All switches in MRCs operate with ZVS, which reduces the switching losses and switching noise and eliminates the oscillation due to the parasitic effects of the components (such as the junction capacitance of the diodes). However, all switches are under high current and voltage stresses, resulting in an increase in the conduction loss.

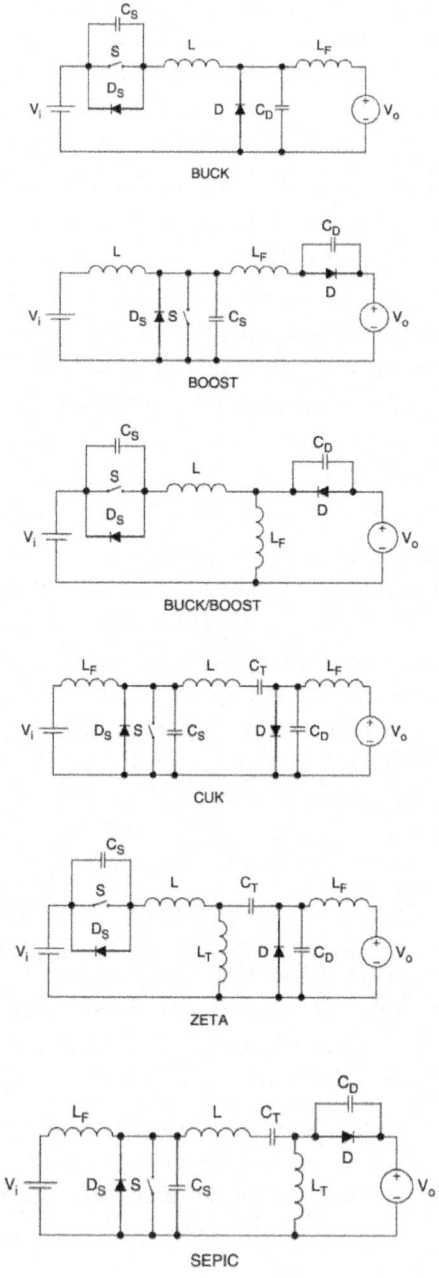

Figure 17: Use of the multi-resonant switch in conventional PWM converters.

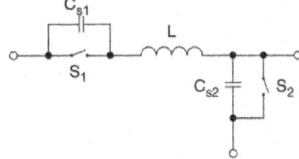

Figure 18: Constant frequency multi-resonant switch.

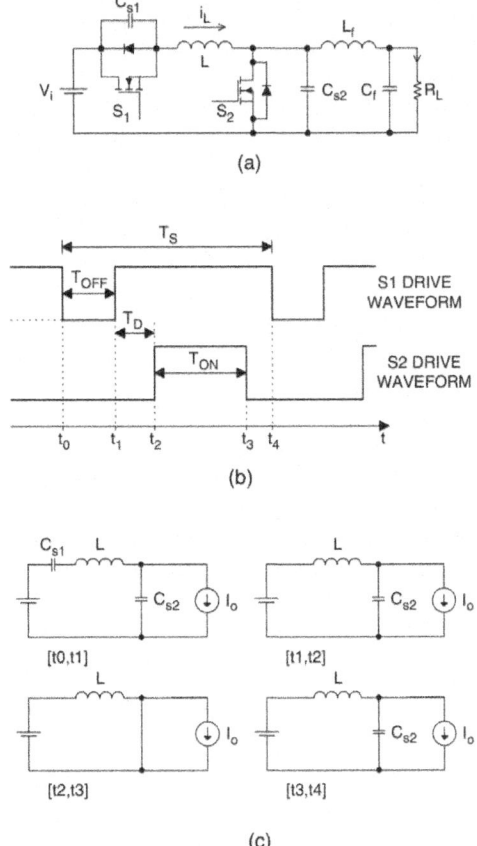

Figure 19: Constant frequency buck MRC: (a) circuit schematics; (b) gate drive waveforms; and (c) operating stages.

ZERO-VOLTAGE-TRANSITION (ZVT) CONVERTERS

By introducing a resonant circuit in parallel with the switches, the converter can achieve ZVS for both power switch and diode without significantly increasing their voltage and currentstresses [23]. Figure 20a shows a buck type ZVT-PWM converter and Fig. 20b shows the associated waveforms. The converter consists of a main switch S and an auxiliary switch S1. It can be seen that the voltage and current waveforms of the switches are square-wave-like except during turn-on and turn-off switching intervals, where ZVT takes place. The main switch and the output diode are under ZVS and are subjected to low voltage and current stresses. The auxiliary switch is under ZCS, resulting in low switching loss.

The concept of ZVT can be extended to other PWM circuits by adding the resonant circuit. Some basic ZVT-PWM converters are shown in Fig. 21.

NON-DISSIPATIVE ACTIVE CLAMP NETWORK

The active-clamp circuit can utilize the transformer leakage inductance energy and can minimize the the turn-off voltage stress in the isolated converters. The active clamp circuit provides a means of achieving ZVS for the power switch and reducing the rate of change of the diode's reverse recovery current. An example of a flyback converter with active clamp is shown in Fig. 22a and the circuit waveforms are shown in Fig. 22b. Clamping action is obtained by using a series combination of an active switch (i.e. S_2) and a large capacitor so that the voltage across the main switch (i.e. S_1) is clamped to a minimum value. S_2 is turned on with ZVS. However, S_2 is turned off with finite voltage and current, and has turnoff switching loss. The clamp-mode ZVS-MRCs is discussed in [24–26].

LOAD RESONANT CONVERTERS

Load resonant converters (LRCs) have many distinct features over conventional power converters. Due to the soft commutation of the switches, no turn-off loss or stress is present. LRCs are specially suitable for high-power applications because they allow high-frequency operation for equipment size/weight reduction, without sacrificing the conversion efficiency and imposing extra stress on the switches. Basically, LRCs can be divided into three different configurations, namely series resonant converters, parallel resonant converters, and series–parallel resonant converters.

Series Resonant Converters

Series resonant converters (SRCs) have their load connected in series with the resonant tank circuit, which is formed by L_r and Cr [15, 27–29]. The half-bridge configuration is shown in Fig. 23. When the resonant inductor current i_{Lr} is positive, it flows through T_1 if T_1 is on; otherwise it flows through the diode D_2. When i_{Lr} is negative, it flows through T_2 if T_2 is on; otherwise it flows through the diode D_1.

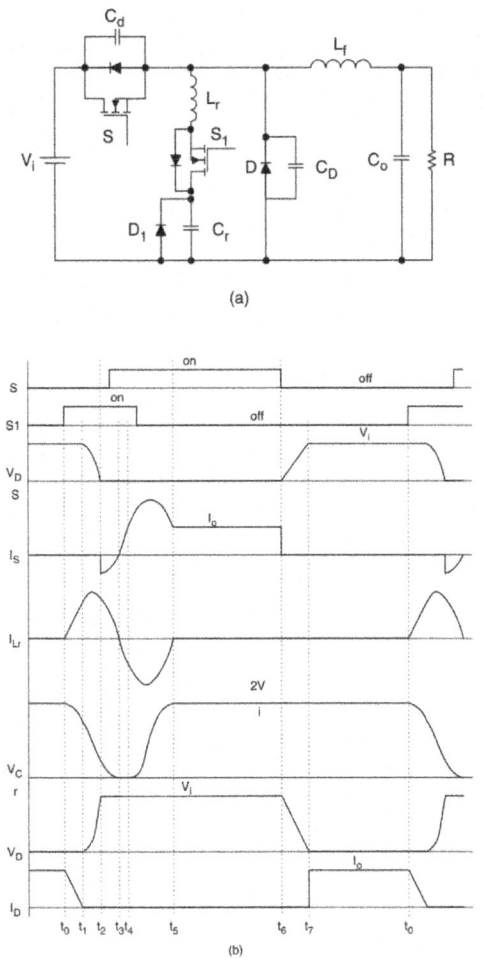

(a)

(b)

Figure 20: Buck ZVT-PWM converter: (a) circuit schematics and (b) waveforms.

In the steady-state symmetrical operation, both the active switches are operated in a complementary manner. Depending on the ratio between the switching frequency ω_S and the converter resonant frequency ω_r, the converter has several possible operating modes.

Discontinuous Conduction Mode (DCM) with $_s < 0.5$ $_r$

Figure 24a shows the waveforms of i_{Lr} and the resonant capacitor voltage v_{Cr} in this mode of operation. From 0 to t_1, T_1 conducts. From t_1 to t_2, the current in T_1 reverses its direction. The current flows through D_1 and back to the supply source. From t_2 to t_3, all switches are in the off state. From t_3 to t_4, T_2 conducts. From t_4 to t_5, the current in T_2 reverses its direction. The current flows through D_2 and back to the supply source. T_1 and T_2 are switched on under ZCS condition and they are switched off under zero-current and zero-voltage conditions. However, the switches are under high current stress in this mode of operation and thus have higher conduction loss.

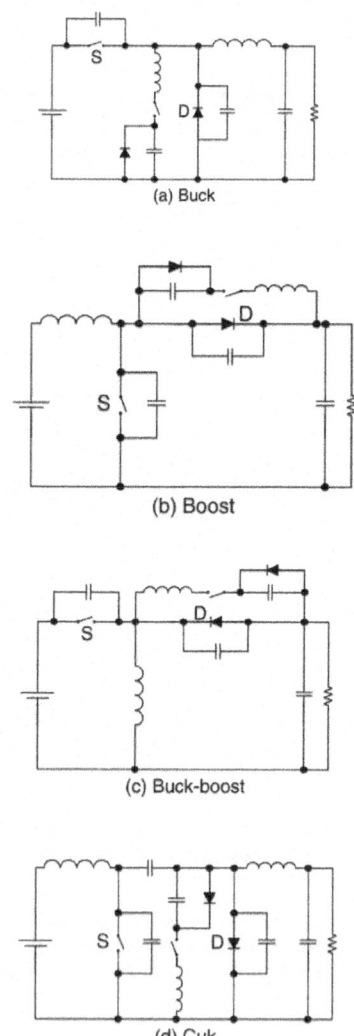

(a) Buck

(b) Boost

(c) Buck-boost

(d) Cuk

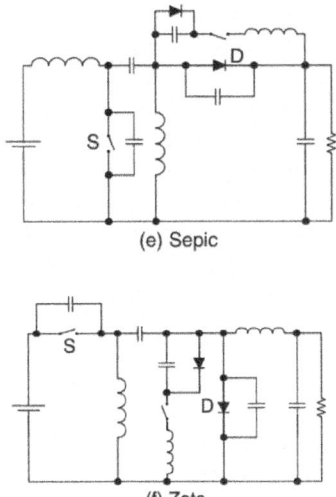

(e) Sepic

(f) Zeta

Figure 21: Conventional ZVT-PWM converters.

Continuous Conduction Mode (CCM) with $0.5\omega_r < \omega_s < \omega_r$

Figure 24b shows the circuit waveforms. From 0 to t_1, i_{Lr} transfers from D_2 to T_1. T_1 is switched on with finite switch current and voltage, resulting in turn-on switching loss. Moreover, the diodes must have good reverse recovery characteristics in order to reduce the reverse recovery current. From t_1 to t_2, D_1 conducts and T_1 is turned off softly with zero voltage and zero current. From t_2 to t_3, T_2 is switched on with finite switch current and voltage. At t_3, T_2 is turned off softly and D_2 conducts until t_4.

Continuous Conduction Mode (CCM) with $\omega_r < \omega_s$

Figure 24c shows the circuit waveforms. From 0 to t_1, iLr transfers from D_1 to T_1. Thus, T_1 is switched on with zerocurrent and zero voltage. At t_1, T_1 is switched off with finite voltage and current, resulting in turn-off switching loss. From t_1 to t_2, D_2 conducts. From t_2 to t_3, T_2 is switched on with zero current and zero voltage. At t_3, T_2 is switched off. i_{Lr} transfers from T_2 to D_1. As the switches are turned on with ZVS, lossless snubber capacitors can be added across the switches.

The following parameters are defined: voltage conversion ratio M, characteristic impedance Z_r, resonant frequency f_r, normalized load resistance r, normalized switching frequency γ.

$$M = nV_o/V_{in} \tag{2a}$$

$$Z_r = \sqrt{L_r/C_r} \tag{2b}$$

$$f_r = 1/\left(2\pi\sqrt{L_r C_r}\right) \tag{2c}$$

$$r = n^2 R_L/Z_r \tag{2d}$$

$$\gamma = f_s/f_r \tag{2e}$$

$$M = 1\left/\sqrt{(\gamma - 1/\gamma)^2/(r^2 + 1)}\right. \tag{2f}$$

The relationships between M and γ for different value of r are shown in Fig. 25. The boundary between CCM and DCM is at r = 1.27γ. When the converter is operating in DCM and 0.2 <γ< 0.5, M = 1.27 rγ.

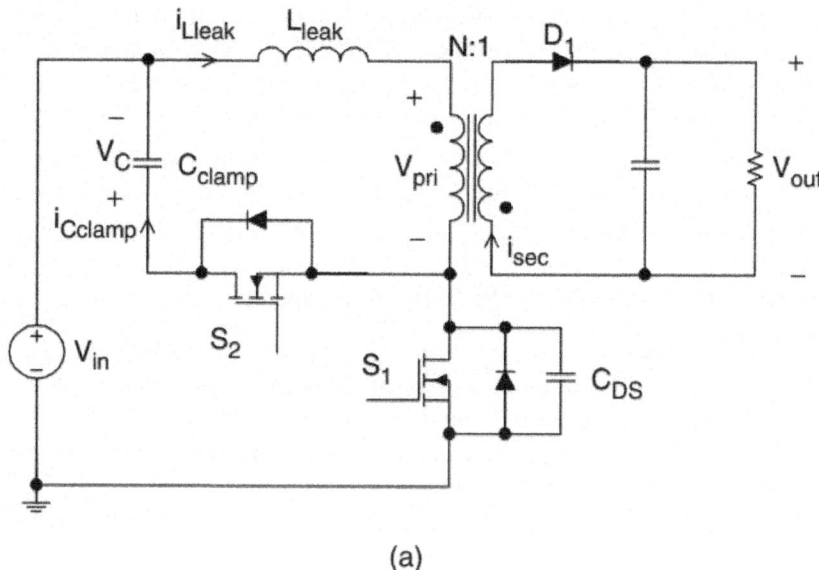

(a)

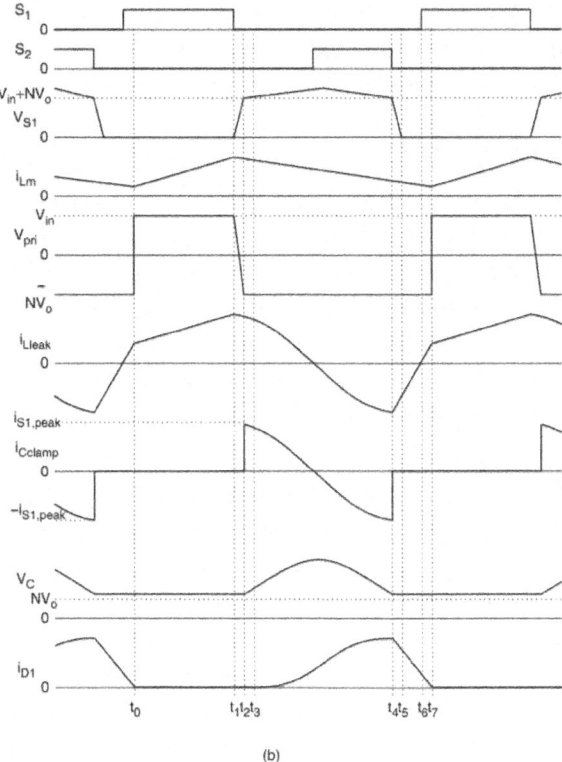

(b)

Figure 22: Active-clamp flyback converter: (a) circuit schematics and (b) circuit waveforms.

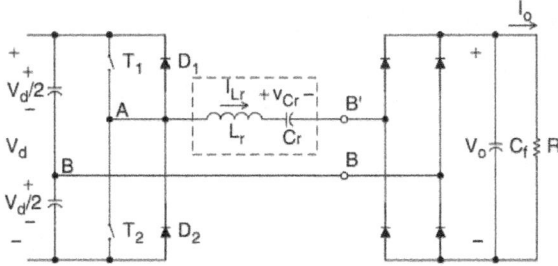

Figure 23: SRC half-bridge configuration.

The SRC has the following advantages. Transformer saturation can be avoided since the series capacitor can block the dc component. The light load efficiency is high because the device current and conduction loss are low. However, the major disadvantages are that there is difficulty in regulating the output voltage under light load and no load conditions. Moreover, the output dc filter capacitor has to carry high ripple current, which could be a major problem in low-output voltage and high-output current applications [29].

Parallel Resonant Converters

Parallel resonant converters (PRCs) have their load connected in parallel with the resonant tank capacitor C_r [27–30]. The half-bridge configuration is shown in Fig. 26. SRC behaves as a current source, whereas the PRC acts as a voltage source. For voltage regulation, PRC requires a smaller operating frequency range than the SRC to compensate for load variation.

Discontinuous Conduction Mode (DCM)

The steady-state waveforms of the resonant inductor current i_{Lr} and the resonant capacitor voltage v_{Cr} are shown in Fig. 27a. Initially both i_{Lr} and v_{Cr} are zero. From 0 to t_2, T_1 conducts and is turned on with zero current. When i_{Lr} is less than the output current I_o, i_{Lr} increases linearly from 0 to t_1 and the output current circulates through the diode bridge. From t_1 to t_3, L_r resonates with C_r. Starting from t_2, i_{Lr} reverses its direction and flows through D_1. T_1 is then turned off with zero current and zero voltage. From t_3 to t_4, v_{Cr} decreases linearly due to the relatively constant value of I_o. At t_4, when v_{Cr} equals zero, the output current circulates through the diode bridge again. Both i_{Lr} and v_{Cr} will stay at zero for an interval. From t_5 to t_9, the above operations will be repeated for T_2 and D_2. The output voltage is controlled by adjusting the time interval of $[t_4, t_5]$.

Continuous Conduction Mode $\omega_s < \omega_r$

This mode is similar to the operation in the DCM, but with a higher switching frequency. Both i_{Lr} and v_{Cr} become continuous. The waveforms are shown in Fig. 27b. The switches T_1 and T_2 are hard turned on with finite voltage and current and are soft turned off with ZVS.

Continuous Conduction Mode $\omega_s > \omega_r$

If the switching frequency is higher than ω_r (Fig. 27c), the anti-parallel diode of the switch will be turned on before the switch is triggered. Thus, the switches are turned on with ZVS. However, the switches are hard turned off with finite current and voltage.

The parameters defined in Eq. (2) are applicable. The relationships between M and γ for various values of r are shown in Fig. 28. During the DCM (i.e.

$\gamma < 0.5$), M is in linear relationship with γ. Output voltage regulation can be achieved easily. The output voltage is independent on the output current. The converter shows a good voltage source characteristics. It is also possible to step up and step down the input voltage.

The PRC has the advantages that the load can be shortcircuited and the circuit is suitable for low-output voltage, high-output current applications. However, the major disadvantage of the PRC is the high device current. Moreover, since the device current do not decrease with the load, the efficiency drops with a decrease in the load [29].

Series–Parallel Resonant Converter

Series–Parallel Resonant Converter (SPRC) combines the advantages of the SRC and PRC. The SPRC has an additional capacitor or inductor connected in the resonant tank circuit [29–31]. Figure 29a shows an LCC-type SPRC, in which an additional capacitor is placed in series with the resonant inductor. Figure 29b shows an LLC-type SPRC, in which an additional inductor is connected in parallel with the resonant capacitor in the SRC. However, there are many possible combinations of the resonant tank circuit. Detailed analysis can be found in [31].

CONTROL CIRCUITS FOR RESONANT CONVERTERS

Since the 1985s, various control integrated circuits (ICs) for resonant converters have been developed. Some common ICs for different converters are described in this section.

QRCs and MRCs

Output regulations in many resonant-type converters, such as QRCs and MRCs, are achieved by controlling theswitching frequency. ZCS applications require controlled switch-on times while ZVS applications require controlled switch-off times. The fundamental control blocks in the IC include an error amplifier, voltage controlled oscillator (VCO), one shot generator with a zero wave-crossing detection comparator, and an output stage to drive the active switch. Typical ICs include UC1861–UC1864 for ZVS applications and UC 1865–UC 1868 for ZCS applications [32]. Figure 30 shows the controller block diagram of UC 1864.

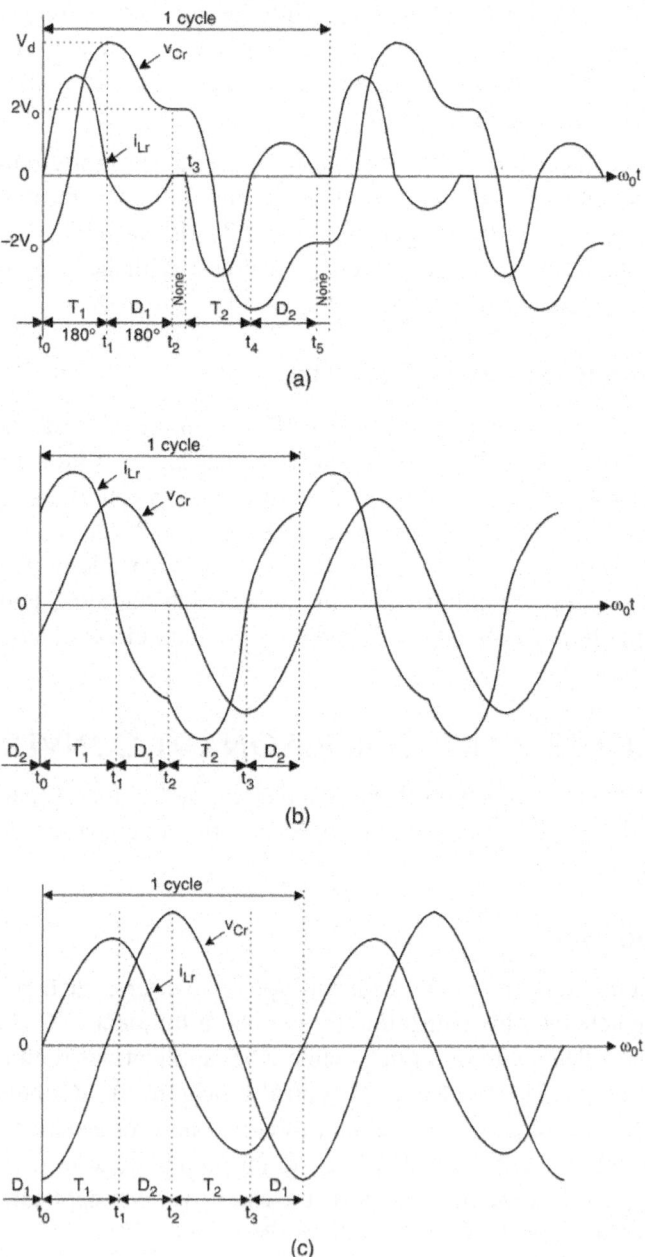

Figure 24: Circuit waveforms under different operating conditions: (a) $\omega_s <$ 0.5 ω_r; (b) 0.5 $\omega_r < \omega_s < \omega$; and (c) $\omega_r < \omega_s$.

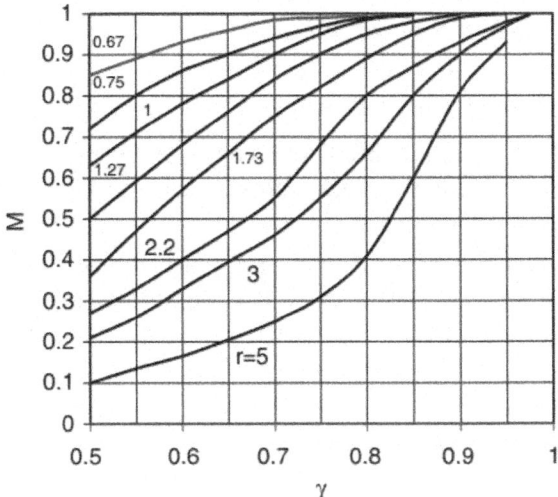

Figure 25: M vs γ in SRC.

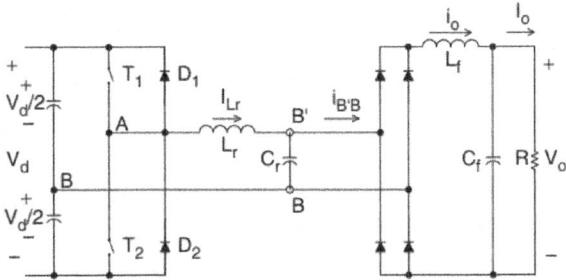

Figure 26: PRC half-bridge configuration.

The maximum and minimum switching frequencies (i.e. f_{max} and f_{min}) are controlled by the resistors Range and R_{min} and the capacitor C_{vco}. f_{max} and f_{min} can be expressed as

$$f_{max} = \frac{3.6}{(R_{ange}//R_{min})C_{VCO}} \quad \text{and} \quad f_{min} = \frac{3.6}{R_{min}C_{VCO}}$$

(3)

The frequency range Δf is then equal to

$$\Delta f = f_{max} - f_{min} = \frac{3.6}{R_{ange}C_{VCO}}$$

(4)

The frequency range of the ICs is from 10 kHz to 1 MHz. The output frequency of the oscillator is controlled by the error amplifier (E/A) output. An example of a ZVS-MR forward converter is shown in Fig. 31.

Phase-Shifted, ZVT FB Circuit

The UCC3895 is a phase shift PWM controller that can generate a phase shifting pattern of one half-bridge with respect to the other. The application diagram is shown in Fig. 32.

The four outputs "OUTA," "OUTB," "OUTC," and "OUTD" are used to drive the MOSFETs in the full-bridge. The dead time between "OUTA" and "OUTB" is controlled by "DELAB" and the dead time between "OUTC" and "OUTD" is controlled by "DELCD." Separate delays are provided for the two half-bridges to accommodate differences in resonant capacitor charging currents. The delay in each set is approximated by

$$t_{DELAY} = \frac{25 \times 10^{-12} R_{DEL}}{0.75(V_{CS} - V_{ADS}) + 0.5} + 25\ ns \tag{5}$$

where R_{DEL} is the resistor value connected between "DELAB" or "DELCD" to ground.

The oscillator period is determined by R_T and C_T. It is defined as

$$t_{OSC} = \frac{5 R_T C_T}{48} + 120\ ns \tag{6}$$

The maximum operating frequency is 1 MHz. The phase shift between the two sets of signals is controlled by the ramp voltage and the error amplifier output having a 7 MHz bandwidth.

EXTENDED-PERIOD QUASI-RESONANT (EP-QR) CONVERTERS

Generally, resonant and quasi-resonant converters operate with frequency control. The extended-period quasi-resonant converters proposed by Barbi [33] offer a simple solution to modify existing hard-switched converters into soft-switched ones with constant frequency operation. This makes both output filter design and control simple. Figure 33 shows a standard hard-switched boost type PFC converter. In this hardswitched circuit, the main switch SW1 could be subject to significant switching stress because the reverse recovery current of the diode D_F could be excessive when SW1 is turned on. In practice, a small saturable inductor may be added in series with the power diode D_F in order to reduce the di/dt of the reverse-recovery current. In addition, an optional R–C snubber may be added across SW1 to reduce the dv/dt of SW1. These extra reactance components can in fact be used in the EP-QR circuit to achieve soft switching, as shown in Fig. 34. The resonant components L_r and C_r are of small values and can come from the snubber circuits of a standard hard-switched converter.

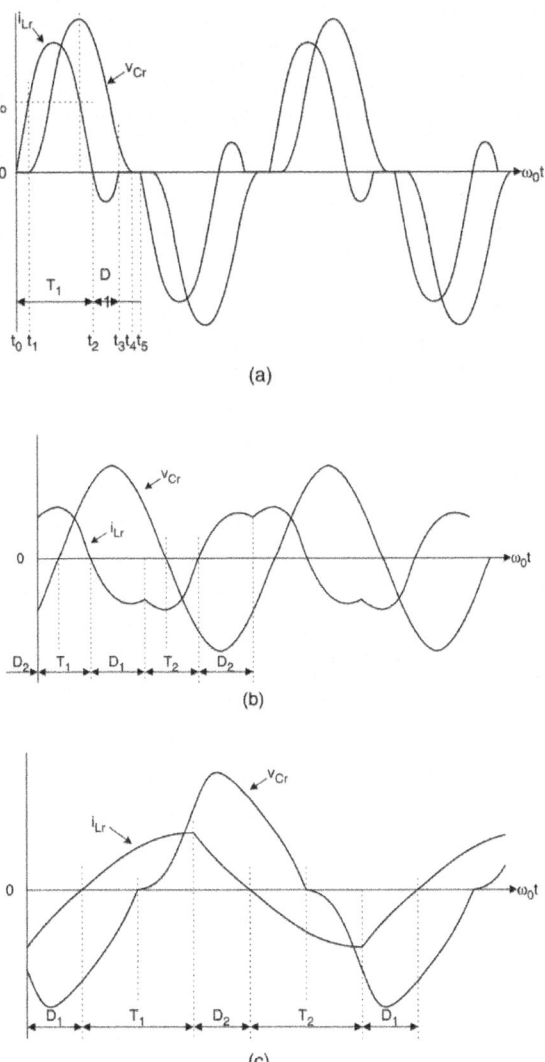

Figure 27: Circuit waveforms under different operating conditions: (a) discontinuous conduction mode; (b) continuous conduction mode $\omega_s < \omega_r$; and (c) continuous conduction mode $\omega_s > \omega_r$.

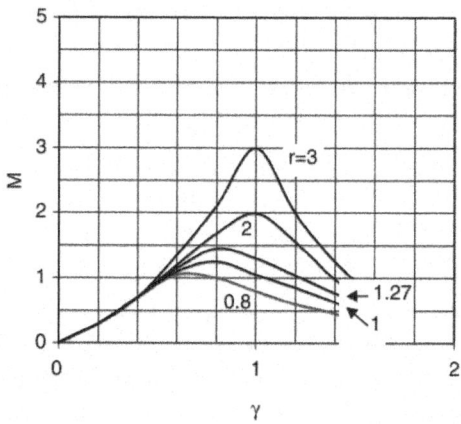

Figure 28: M vs γ in PRC.

Thus, the only additional component is the auxiliary switch Q2. The small resonant inductor is put in series with the main switch SW1 so that SW1 can be switched on under ZC condition and the di/dt problem of the reverserecovery current be eliminated. The resonant capacitor C_r is used to store energy for creating condition for soft switching. Q2 is used to control the resonance during the main switch transition. It should be noted that all power devices including SW1, Q1 and main power diode D_F are turned on and off under ZV and/or ZC conditions. Therefore, the large di/dt problem due to the reverse recovery of the power diode can be eliminated. The soft-switching method is an effective technique for EMI suppression.

Together with power factor correction technique, softswitching converters offer a complete solution to meet EMI regulations for both conducted and radiated EMI. The operation of the EP-QR boost PFC circuit [34, 35] can be described in six modes as shown in Fig. 35. The corresponding idealized waveforms are included in Fig. 36.

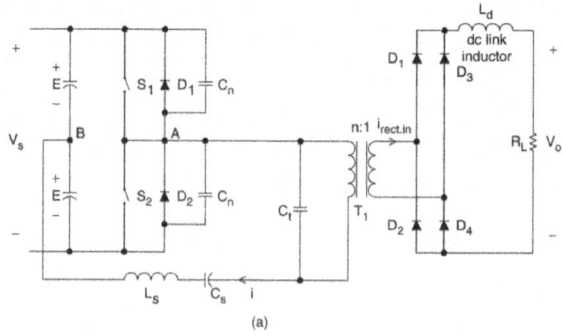

(a)

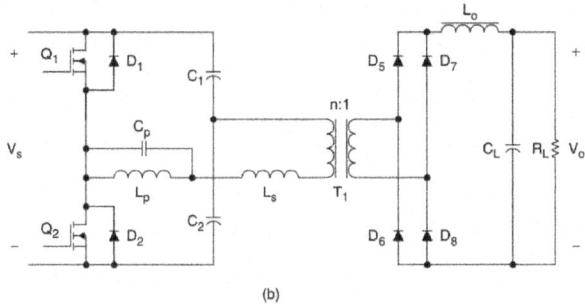

(b)

Figure 29: Different types of SPRC: (a) LCC-type and (b) LLC-type.

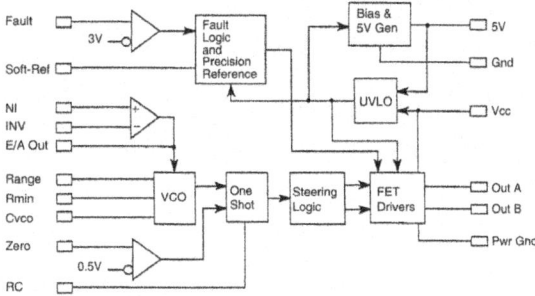

Figure 30: Controller block diagram of UC1864 (Courtesy of Unitrode Corp. and Texas Instruments).

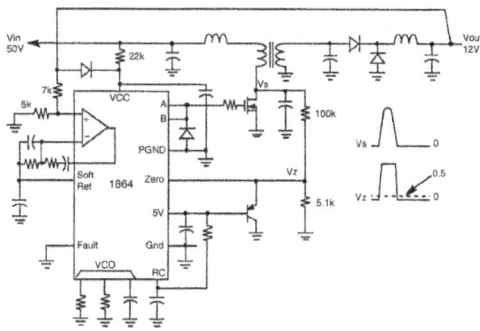

Figure 31: ZV-MR forward converter (Courtesy of Unitrode Corp. and Texas Instruments).

Circuit Operation

Interval I: (t_0-t_1) Due to the resonant inductor L_r which limits the di/dt of the switch current, switch SW1 is turned on at zero-current condition with a positive gating signal V_{GS1} to start a switching cycle at $t = t_0$. Current in D_F is diverted to inductor L_r. Because D_F is still conducting during this short period, D_{S2} is still reverse biased and is thus not conducting. The equivalent circuit topology for the conducting paths is shown in Fig. 35a. Resonant switch Q_2 remains off in this interval.

Interval II: (t_1-t_a) When D_F regains its blocking state, D_{S2} becomes forward biased. The first half of the resonance cycle occurs and resonant capacitor C_r starts to discharge and current flows in the loop $C_r-Q_2-L_r-SW_1$. The resonance half-cycle stops at time $t = t_a$ because D_{S2} prevents the loopcurrent i_{Cr} from flowing in the opposition direction. The voltage across C_r is reversed at the end of this interval. The equivalent circuit is shown in Fig. 35b.

Interval III: (t_a-t_b) Between t_a and t_b, current in L_F and L_r continues to build up. This interval is the extended-period for the resonance during which energy is pumped into L_r. The corresponding equivalent circuit is showed in Fig. 35c.

Interval IV: (t_b-t_2) Figure 35d shows the equivalent circuit for this operating mode. Before SW_1 is turned off, the second half of the resonant cycle needs to take place in order that a zero-voltage condition can be created for the turn-off process of SW_1. The second half of the resonant cycle starts when auxiliary switch Q_2 is turned on at $t = t_b$. Resonant current then flows through the loop $L_r-Q_2-C_r$-anti-parallel diode of SW_1.

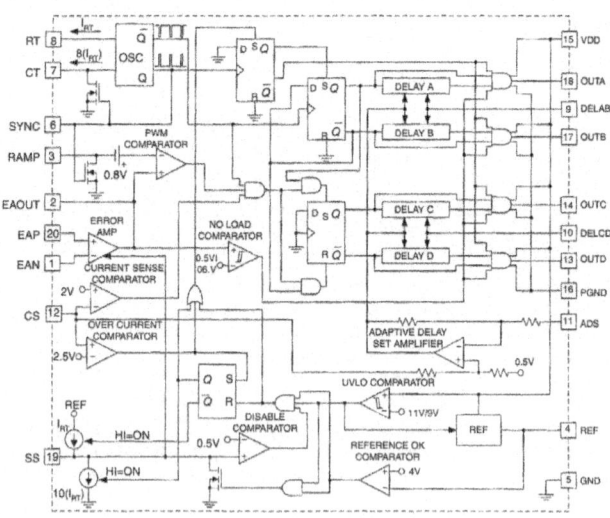

Figure 32: Application diagram of UCC3895 (Courtesy of Unitrode Corp. and Texas Instruments).

This current is limited by L_r and thus Q_2 is turned on under zero-current condition. Since the anti-parallel diode of SW_1 is conducting, the voltage across SW_1 is clamped to the on-state voltage of the anti-parallel diode. SW_1 can therefore be turned off at (near) zero-voltage condition before $t = t_2$ at which the second half of the resonant cycle ends.

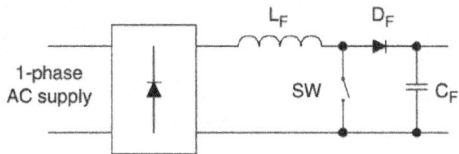

Figure 33: Boost-type AC–DC power factor correction circuit.

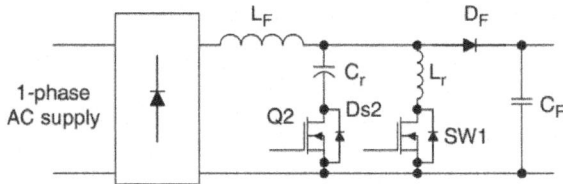

Figure 34: EP-QR boost-type AC–DC power factor correction circuit.

Interval V: (t_2-t_3) During this interval, the voltage across C_r is less than the output voltage V_o. Therefore D_F is still reverse biased. Inductor current I_s flows into C_r until V_{Cr} reaches V_o at $t = t_3$. The equivalent circuit is represented in Fig. 35e.

Interval VI: (t_3-t_4) During this period, the resonant circuit is not in action and the inductor current Is charges the output capacitor C_F via D_F, as in the case of a classical boost-type PFC circuit. C_r is charged to V_o, therefore Q_2 can be turned off at zero-voltage and zero-current conditions. Figure 35f shows the equivalent topology of this operating mode.

In summary, SW_1, Q_2, and D_F are fully soft-switched. Since the two resonance half-cycles take place within a closed loop outside the main inductor, the high resonant pulse will not occur in the inductor current, thus making a well-established averaged current mode control technique applicable for such QR circuit. For full soft-switching in the turn-off process, the resonant components need to be designed so that the peak resonant current exceeds the maximum value of the inductor current. Typical measured switching waveforms and trajectories of SW_1, Q_2 and D_F are shown in Figs. 37, 38, and 39, respectively.

Design Procedure

Given: Input AC voltage = V_s (V)

Peak AC voltage = $V_s(\text{max})$ (V)

Nominal output DC voltage = V_o (V)

Switching frequency = f_{sw} (Hz)

Output power = P_o (W)

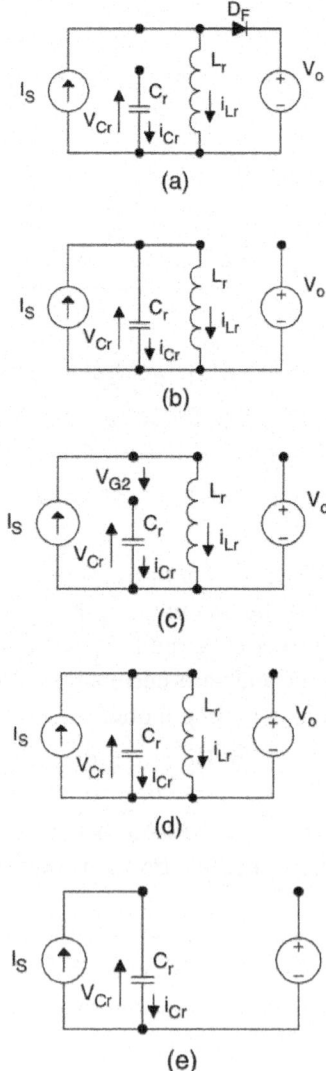

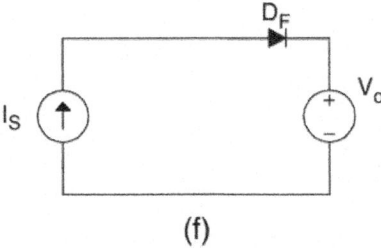

(f)

Figure 35: Operating modes of EP-QR boost-type AC–DC power factor correction circuit.

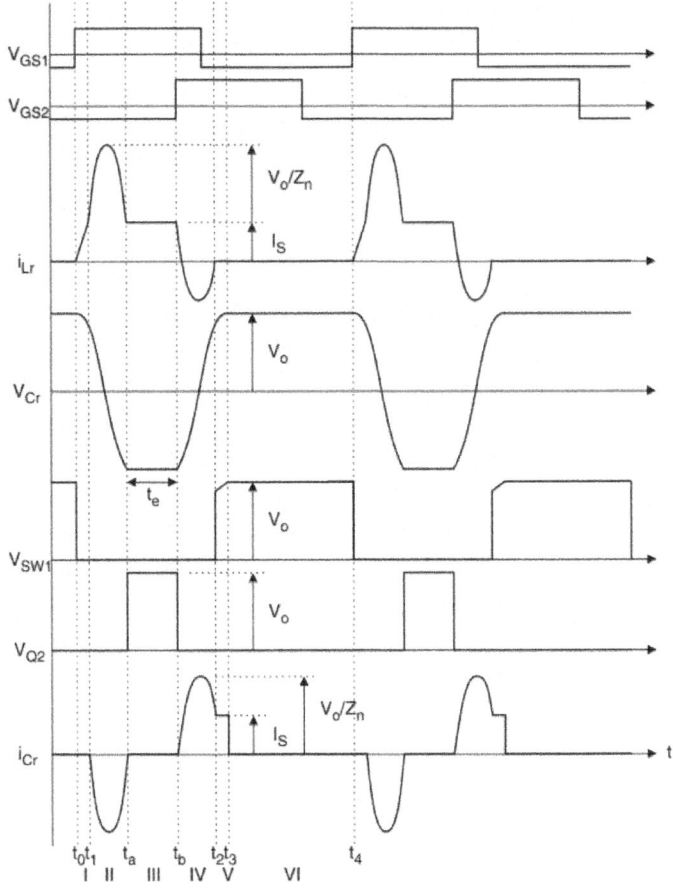

Figure 36: Idealized waveforms of EP-QR boost-type AC–DC power factor correction circuit.

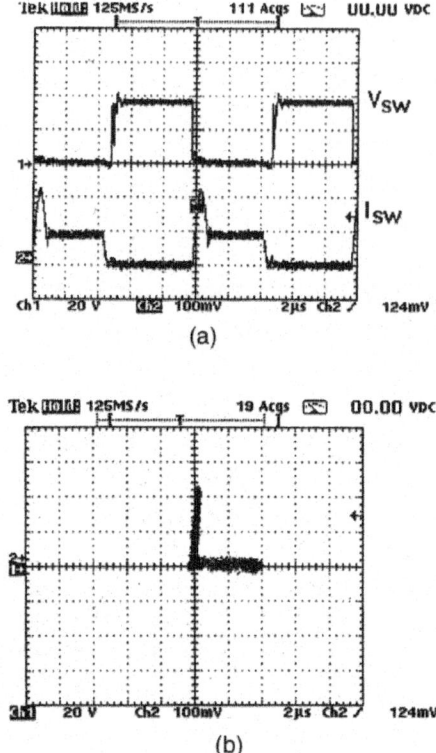

Figure 37: (a) Drain-source voltage and current of SW$_1$ and (b) switching locus of SW$_1$.

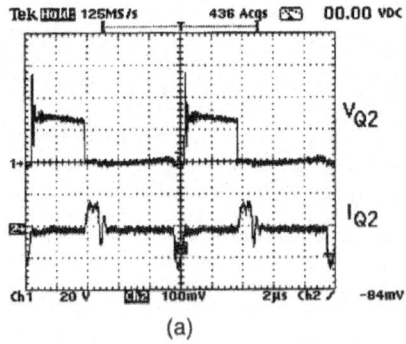

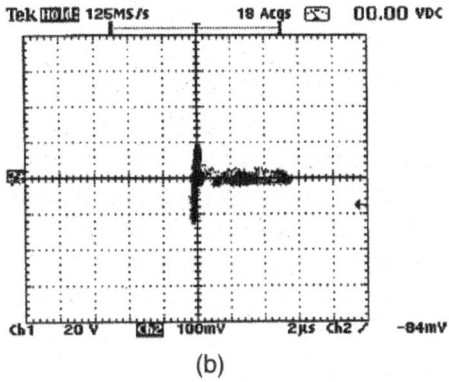

(b)

Figure 38: (a) Drain-source voltage and current of Q_2 and (b) switching locus of Q_2.

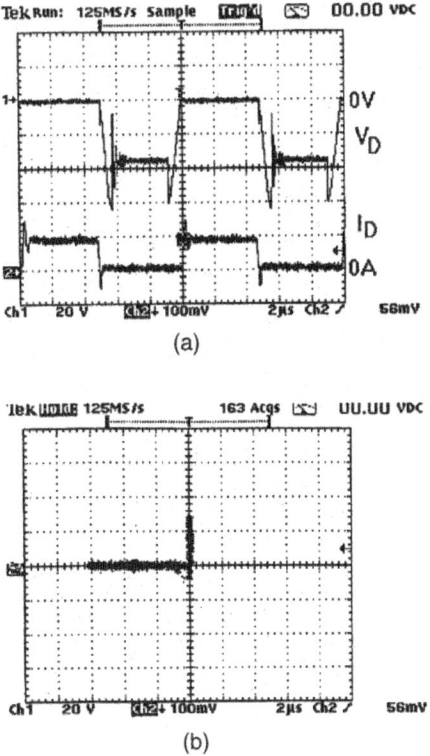

Figure 39: (a) Diode voltage and current and (b) switching locus of diode.

Input current ripple = ΔI (A)

Output voltage ripple = ΔV (V)

(I) Resonant tank design:

Step 1: Because the peak resonant current must be greater than the peak inductor current (same as peak input line current) in order to achieve soft-turn-off, it is necessary to determine the peak input current $I_s(max)$. Assuming lossless AC–DC power conversion, $I_s(max)$ can be estimated from the following equation

$$I_{s(max)} \approx \frac{2V_o I_o}{V_{s(max)}}$$

(7)

where $I_o = P_o/V_o$ is the maximum output current.

Step 2: Soft-switching criterion is

$$Z_r \leq \frac{V_o}{I_{s(max)}}$$

(8)

where $Z_r = \sqrt{\frac{L_r}{C_r}}$ is the impedance of the resonant tank

For a chosen resonant frequency f_r, L_r, and C_r can be obtained from:

$$2\pi f_r = \frac{1}{\sqrt{L_r C_r}}$$

(9)

(II) Filter component design:

The minimum conversion ratio is

$$M_{(min)} = \frac{V_o}{V_{s(max)}}$$

$$= \frac{1}{1 - \left(\frac{f_{sw}}{f_r} + \frac{t_e}{T_{sw}}\right)}$$

(10)

where $T_{sw} = 1/f_{sw}$ and t_e is the extended period. From Eq. (10), minimum t_e can be estimated.

The turn-on period of the SW_1 is

$$T_{on}(sw1) = t_e + 1/f_r \qquad (11)$$

Inductor value L is obtained from:

$$L \geq \left(\frac{T_{on(sw1)}}{\Delta I} \right) V_{s(max)} \qquad (12)$$

The filter capacitor value C can be determined from:

$$C \left(\frac{\Delta V}{\frac{T_s}{\pi} \sin^{-1} \left(\frac{I_o}{I_{s(max)}} \right)} \right) = I_o \qquad (13)$$

where $T_s = 1/f_s$ is the period of the AC voltage supply frequency.

Soft-switched DC–DC Flyback Converter

A simple approach that can turn an existing hard-switched converter design into a soft-switched one is shown in Fig. 40. The key advantage of the proposal is that many well proven and reliable hard-switched converter designs can be kept. The modification required is the addition of a simple circuit (consisting an auxiliary winding, a switch, and a small capacitor) to an existing isolated converter [36]. This principle, which is the modified version of the EPQR technique for isolated converter, is demonstrated in an isolated soft-switched flyback converter with multiple outputs. Other advantageous features of the proposal are:

- All switches and diodes of the converter are 'fully' soft-switched, i.e. soft-switched at both turn-on and turnoff transitions under zero-voltage and/or zero-current conditions.

- The leakage inductance of each winding in the flyback transformer forms part of the resonant circuit for achieving ZVS and ZCS of all switches and diodes.

- The control technique is simply PWM-based as in standard hard-switched converters.

- The soft-switched technique is a proven method for EMI reduction [37].

A ZCS Bidirectional Flyback DC–DC Converter

A bi-directional flyback dc–dc converter that uses one auxiliary circuit for both power flow directions is proposed in Fig. 41 [38]. The methodology is based on extending the unidirectional soft-switched flyback converter in [36] and replacing the output diode with a controlled switch, which acts as either a rectifier [39] or a power control switch in the corresponding power flow direction. An auxiliary circuit that consists of a winding in the coupled inductor, a switch, and a capacitor converts the hard-switched design into a soft-switched one. The operation is the same as [36] in the forward mode. An extended-period resonant stage [34] is introduced when the power control switch is on. Conversely, in the reverse mode, a complete resonant stage is initiated before the main switch is off. In both the power flow operations, the leakage inductance of the coupled inductor is used to create zero-current switching conditions for all switches.

SOFT-SWITCHING AND EMI SUPPRESSION

A family of EP-QR converters are displayed in Fig. 42. Their radiated EMI emission have been compared with that from their hard-switched counterparts [37]. Figures 43a, b show the conducted EMI emission from a hard-switched fly back converter and a soft-switched one, respectively. Their radiated EMI emissions are included in Fig. 44. Both converters are tested at an output power of 50W. No special filtering or shielding measures have been taken during the measurement. It is clear from the measurements that soft-switching is an effective means to EMI suppression.

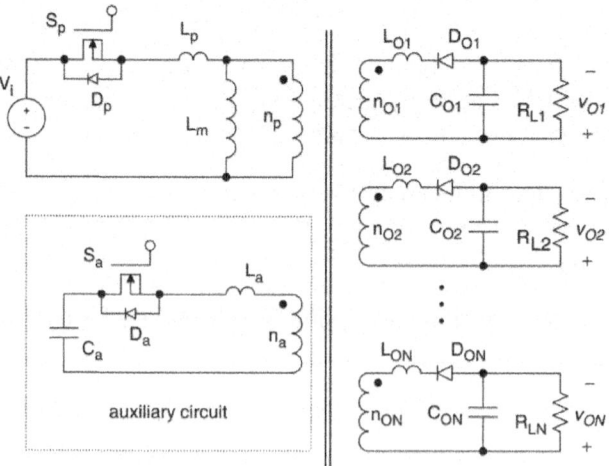

Figure 40: Fully soft-switched isolated flyback converter.

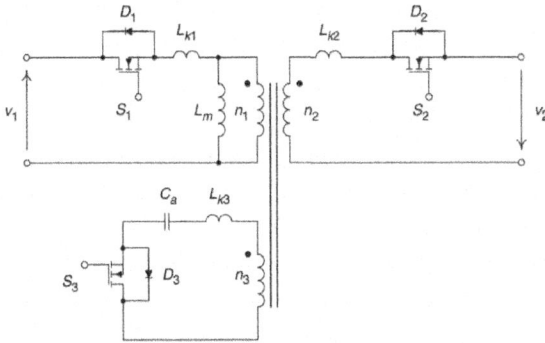

Figure 41: Bidirectional soft-switched isolated flyback converter.

SNUBBERS AND SOFT-SWITCHING FOR HIGH POWER DEVICES

Today, most of the medium power (up to 200 kVA) and medium voltage (up to 800V) inverter are hard-switched. Compared with low-power switched mode power supplies, the high voltage involved in the power inverters makes the dv/dt, di/dt, and the switching stress problems more serious. In addition, the reverse recovery of power diodes in the inverter leg may cause very sharp current spike, leading to severe EMI problem. It should be noted that some high power devices such as GTO thyristors do not have a square safe operating area (SOA). It is therefore essential that the switching stress they undergone must be within their limits. Commonly used protective measures are to use snubber circuits for protecting high power devices.

Among various snubbers, two snubber circuits are most well-known for applications in power inverters. They are the Undeland snubber [40] (Fig. 45) and McMurray snubber circuits [41] (Fig. 46). The Undeland snubber is an asymmetric snubber circuit with one turn-on inductor and one turn-off capacitor. The turn-off snubber capacitor C_s is clamped by another capacitor C_c. At the end of each switching cycle, the snubber energy is dumped into C_c and then discharged into the dc bus via a discharge resistor.

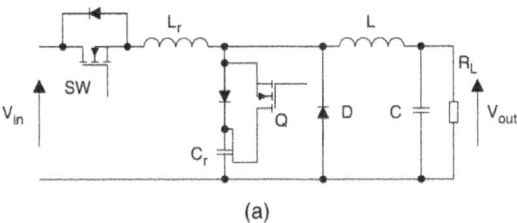

(a)

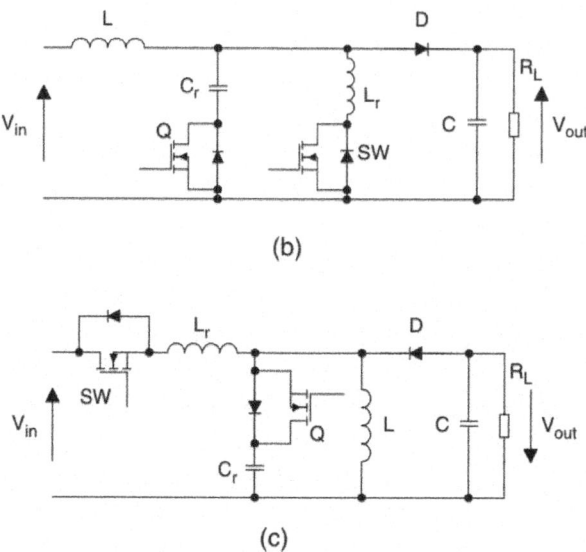

(b)

(c)

Figure 42: A family of EP-QR converters: (a) buck converter; (b) boost converter; and (c) flyback converter.

In order to reduce the snubber loss, the discharge resistor can be replaced by a switched mode circuit. In this way, the Undeland snubber can become a snubber with energy recovery. The McMurray snubber is symmetrical. Both the turn-off snubber capacitors share current in parallel during turn off. The voltage transient is limited by the capacitor closest to the turning-off device because the stray inductance to the other capacitor will prevent instantaneous current sharing. The turn-on inductors require mid-point connection. Snubber energy is dissipated into the snubber resistor. Like the Undeland snubber, the McMurray snubber can be modified into an energy recovery snubber. By using an energy recovery transformer as shown in Fig. 47, this snubber becomes a regenerative one. Although other regenerative circuits have been proposed, their complexity makes them unattractive in industrial applications. Also, they do not necessarily solve the power diode reverse-recovery problems.

Although the use of snubber circuits can reduce the switching stress in the power devices, the switching loss is actually damped into the snubber resistors unless regenerative snubbers are used. The switching loss is still a limiting factor to the high frequency operation of power inverters. However, the advent of soft-switching techniques opens a new way to highfrequency inverter operation. Because the switching trajectory of a soft-switched switch is close to the voltage and current axis, faster power electronic devices with smaller

SOAs can in principle be used. In general, both ZVS and ZCS can reduce switching loss in high-power power switches. However, for power switches with tail currents, such as IGBT, ZCS is more effective than ZVS.

SOFT-SWITCHING DC–AC POWER INVERTERS

Soft-switching technique not only offers a reduction in switching loss and thermal requirement, but also allows the possibility of high frequency and snubberless operation. Improved circuit performance and efficiency, and reduction of EMI emission can be achieved. For zero-voltage switching (ZVS) inverter applications, two major approaches which enable inverters to be soft-switched have been proposed.

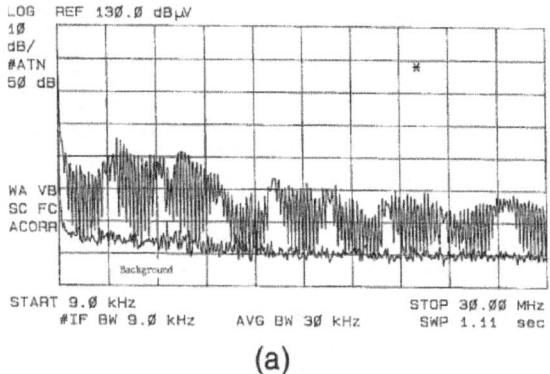

(a)

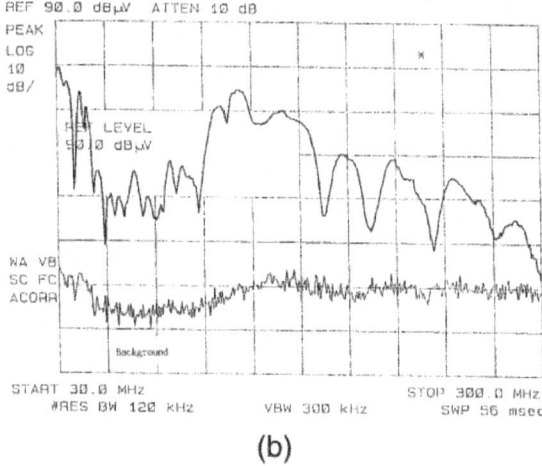

(b)

Figure 43: (a) Conducted EMI from hard-switched flyback converter and (b) radiated EMI from hard-switched flyback converter.

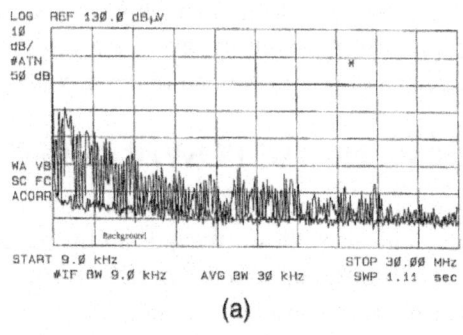

(a)

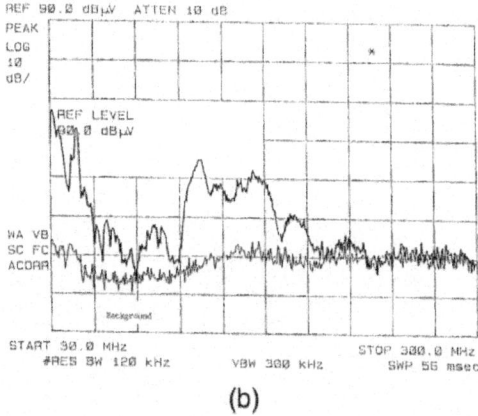

(b)

Figure 44: (a) Conducted EMI from soft-switched flyback converter and (b) radiated EMI from soft-switched flyback converter.

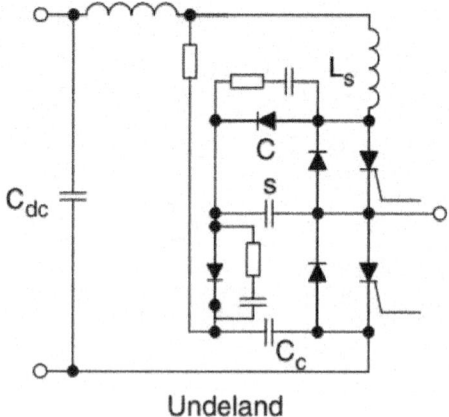

Undeland

Figure 45: Undeland snubber.

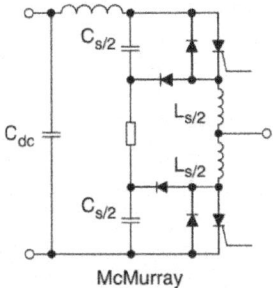

Figure 46: McMurray snubber.

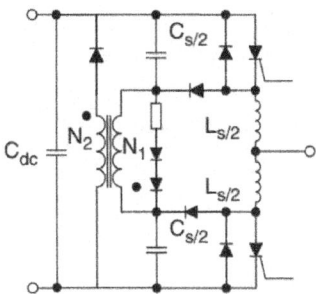

Figure 47: McMurray snubber with energy recovery.

The first approach pulls the dc link voltage to zero momentarily so that the inverter's switches can be turned on and off with ZVS. Resonant dc link and quasi-resonant inverters belong to this category. The second approach uses the resonant pole idea. By incorporating the filter components into the inverter operation, resonance condition and thus zero voltage/current conditions can be created for the inverter switches.

The following soft-switched inverters are described.

Approach 1: Resonant dc link inverters

1. Resonant (pulsating) dc link inverters

2. Actively-clamped resonant dc link inverters

3. Resonant inverters with minimum voltage stress

4. Quasi-resonant soft-switched inverter

5. Parallel resonant dc link inverter.

Approach 2: Resonant pole inverters

1. Resonant pole inverters

2. Auxiliary resonant pole inverters

3. Auxiliary resonant commutated pole inverters.

Type 1 is the resonant dc link inverter [42–44] which sets the dc link voltage into oscillation so that the zero-voltage instants are created periodically for ZVS. Despite the potential advantages that this soft-switching approach can offer, a recent review on existing resonant link topologies for inverters [45] concludes that the resonant dc link system results in an increase in circuit complexity and the frequency spectrum is restricted by the need of using integral pulse density modulation (IPDM) when compared with a standard hard-switched inverter. In addition, the peak pulsating link voltage of resonant link inverters is twice the dc link voltage in a standard hard-switched inverter. Although clamp circuits (Type 2) can be used to limit the peak voltage to 1.3–1.5 per unit [44], power devices with higher than normal voltage ratings have to be used.

Circuits of Type 3–5 employ a switched mode front stage circuit which pulls the dc link voltage to zero momentarily whenever inverter switching is required. This soft-switching approach does not cause extra voltage stress to the inverter and hence the voltage rating of the power devices is only 1 per unit. As ZVS conditions can be created at any time, there is virtually no restriction in the PWM strategies. Therefore, well established PWM schemes developed in the last two decades can be employed. In some ways, this approach is similar to some dc-side commutation techniques proposed in the past for thyristor inverters [46, 47], although these dc-side commutation techniques were used for turning off thyristors in the inverter bridge and were not primarily developed for soft-switching.

Circuits of Type 6–8 retain the use of a constant dc link voltage. They incorporate the use of the resonant components and/or filter components into the inverter circuit operation. This approach is particularly useful for inverter applications in which output filters are required. Examples include uninterruptible power supplies (UPS) and inverters with output filters for motor drives. The LC filter components can form the auxiliary resonant circuits that create the soft-switching conditions. However, these tend to have high power device count and require complex control strategy.

Resonant (Pulsating) DC Link Inverter

Resonant DC link converter for DC–AC power conversion was proposed in 1986 [42]. Instead of using a nominally constant DC link voltage, a resonant circuit is added to cause the DC link voltage to be pulsating at a high frequency. This resonant circuit theoretically creates periodic zero-voltage duration at which the inverter switches can be turned on or off. Figure 48 shows the schematics of the pulsating link inverter. Typical dc link voltage, inverter's phase voltage and the line voltages are shown in Fig. 49. Because the inverter switching can only occur at zero voltage duration, integral pulse density modulation (IPDM) has to be adopted in the switching strategy.

Analysis of the resonant dc link converter can be simplified by considering that the inverter system is highly inductive. The equivalent circuit is shown in Fig. 50.

The link current I_x may vary with the changing load condition, but can be considered constant during the short resonant cycle. If switch S is turned on when the inductor current is I_{Lo}, the resonant dc link voltage can be expressed as

$$V_c(t) = V_s + e^{-\alpha t} \left[-V_s \cos(\omega t) + \omega L I_M \sin(\omega t) \right] \tag{14}$$

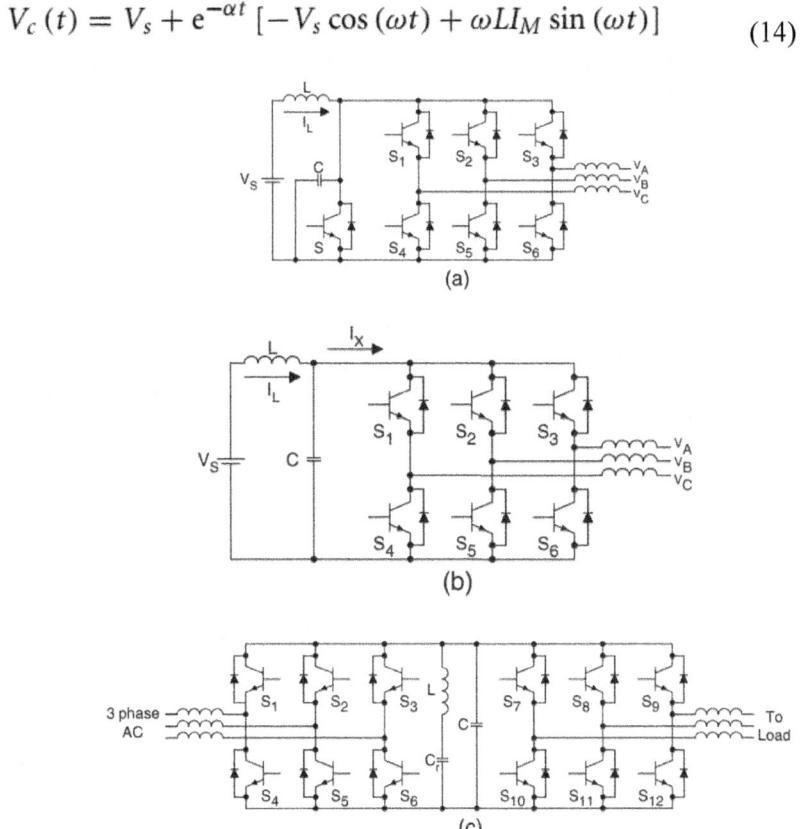

Figure 48: Resonant-link inverters.

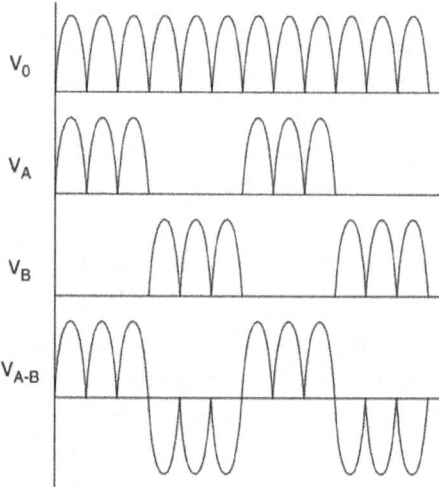

Figure 49: Typical dc link voltage (V_0), phase voltages (V_A, V_B), and line voltage (V_{AB}) of resonant link inverters.

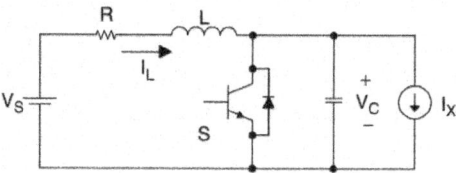

Figure 50: Equivalent circuit of resonant link inverter.

and inductor current i_L is

$$i_L(t) \approx I_x + e^{-\alpha t} \left[I_M \cos(\omega t) + \frac{V_s}{\omega L} \sin(\omega t) \right]$$

(15)

where $\alpha = \dfrac{R}{2L}$

(16)

$$\omega_o = (LC)^{-0.5}$$

(17)

$$\omega = \left(\omega_o^2 - \alpha^2 \right)^{0.5}$$

(18)

and $I_M = I_{Lo} - I_x$

(19)

The resistance in the inductor could affect the resonant behavior because it dissipates some energy. In practice, $(i_L - I_x)$ has to be monitored when S is conducting. S can be turned on when $(i_L - I_x)$ equal to a desired value. The objective is to ensure that the dc link voltage can be resonated to zero voltage level in the next cycle.

The pulsating dc link inverter has the following advantages:

- Reduction of switching loss.

- Snubberless operation.

- High switching frequency (> 18 kHz) operation becomes possible, leading to the reduction of acoustic noise in inverter equipment.

- Reduction of heatsink requirements and thus improvement of power density.

This approach has the following limitations:

- The peak dc pulsating link voltage (2.0 per unit) is higher than the nominal dc voltage value of a conventional inverter. This implies that power devices and circuit components of higher voltage ratings must be used. This could be a serious drawback because power components of higher voltage ratings are not only more expensive, but usually have inferior switching performance than their low-voltage counterparts.

- Although voltage clamp can be used to reduce the peak dc link voltage, the peak voltage value is still higher than normal and the additional clamping circuit makes the control more complicated.

- Integral pulse-density modulation has to be used. Many well-established PWM techniques cannot be employed.

Despite these advantages, this resonant converter concept has paved the way for other soft-switched converters to develop.

Active-clamped Resonant DC Link Inverter

In order to solve the high voltage requirement in the basic pulsating dc link inverters, active clamping techniques (Fig. 51) have been proposed. The active clamp can reduce the perunit peak voltage from 2.0 to about 1.3–1.5 [44]. It has been reported that operating frequency in the range of 60–100 kHz has been achieved [48] with an energy efficiency of 97% for a 50 kVA drive system.

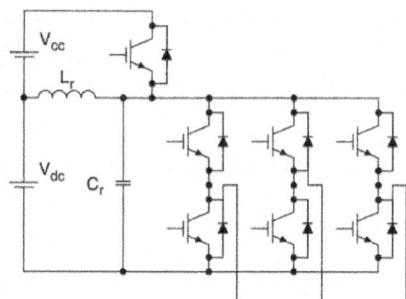

Figure 51: Active-clamp resonant link inverter.

$$T_L = \frac{1}{f_L} = 2\sqrt{L_r C_r}\left(\cos^{-1}(1-k) + \frac{\sqrt{k(2-k)}}{k-1}\right)$$

(20)

where T_L is the minimum link period, f_L is the maximum link frequency, and k is the clamping ratio. For the active-clamped resonant inverter, k is typically 1.3–1.4 per unit.

The rate of rise of the current in the clamping device is

$$\frac{di}{dt} = \frac{(k-1)V_s}{L_r}$$

(21)

The peak clamping current required to ensure that the dc bus return to zero volt is

$$I_{sp} = V_s\sqrt{\frac{k(2-k)C_r}{L_r}}$$

(22)

In summary, resonant (pulsating) dc link inverters offer significant advantages such as:

- High switching frequency operation.

- Low dv/dt for power devices.

- ZVS with reduced switching loss.

- Suitable for 1–250 kW.

- Rugged operation with few failure mode.

Resonant DC Link Inverter with Low Voltage Stress [49]

A resonant dc link inverter with low voltage stress is shown in Fig. 52. It consists of a front-end resonant converter that can pull the dc link voltage down just before any inverter switching. This resonant dc circuit serves as an interface between the dc power supply and the inverter. It essentially retains all the advantages of the resonant (pulsating) dc link inverters. But it offers extra advantages such as:

- No increase in the dc link voltage when compared with conventional hard-switched inverter. That is, the dc link voltage is 1.0 per unit.

- The zero voltage condition can be created at any time. The ZVS is not restricted to the periodic zero-voltage instants as in resonant dc link inverter.

- Well-established PWM techniques can be employed.

- Power devices of standard voltage ratings can be used.

The timing program and the six operating modes of this resonant circuit are as shown in Figs. 53 and 54, respectively

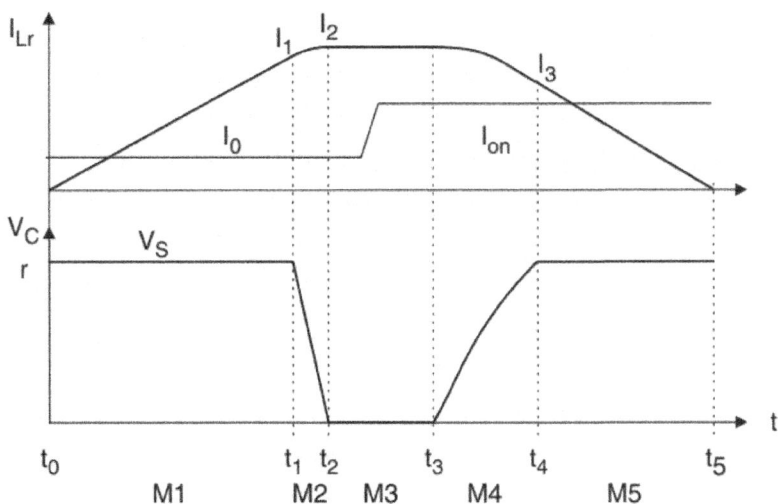

Figure 53: Timing diagram of resonant link inverter with minimum voltage stress.

(1) Normal mode:

This is the standard PWM inverter mode. The resonant inductor current $i_{Lr}(t)$ and the resonant voltage $V_{cr}(t)$ are given by

$$i_{Lr}(t) = 0$$

$$v_{Cr}(t) = V_s$$

where V_s is the nominal dc link voltage.

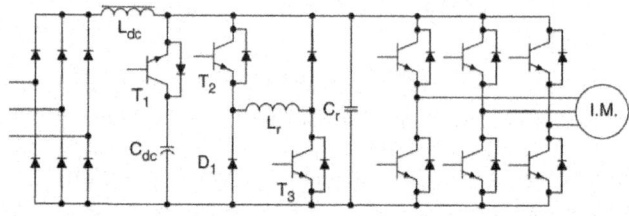

Figure 52: Resonant DC link inverter with low voltage stress.

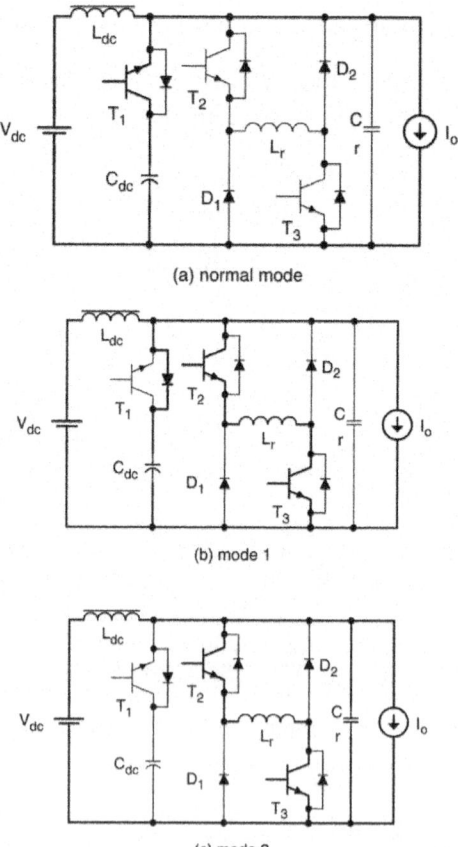

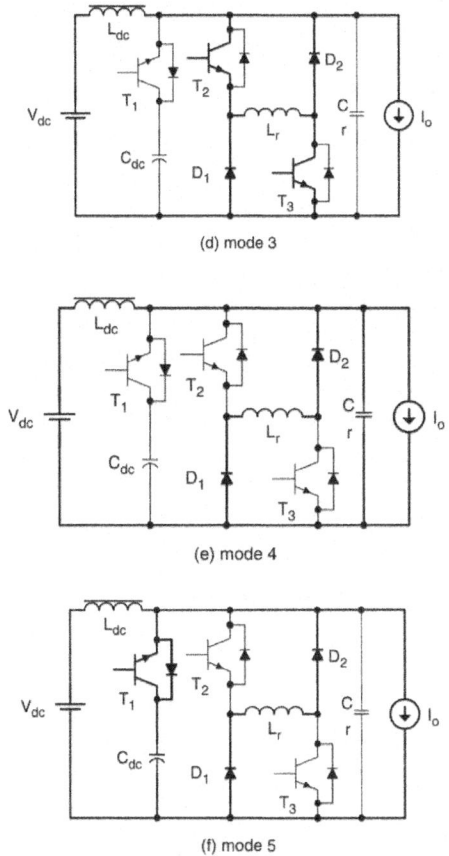

(d) mode 3

(e) mode 4

(f) mode 5

Figure 54: Operating modes of resonant link inverter with minimum voltage stress.

(2) Mode 1 (initiating mode): (t_0-t_1)

At t_0, mode 1 begins by switching on T_2 and T_3 on with zero current. $i_{Lr}(t)$ increases linearly with a di/dt of V_s/L_r. If $i_{Lr}(t)$ is equal to the initialized current I_i, T_1 is zero-voltage turned off. If $(I_s-I_o) < I_i$, then the initialization is ended when $i_{Lr}(t)$ is equal to I_i, where I_s is the current flowing into the dc inductor L_{dc}. If $(I_s-I_o) > I_i$, then this mode continues until $i_{Lr}(t)$ is equal to (I_s-I_o). The equations in this interval are

$$i_{Lr}(t) = \frac{V_s}{L_r}t$$

$$v_{Cr}(t) = V_s$$

$$i_{Lr}(t_1) = \frac{V_s}{L_r}t_1 = I_i$$

(3) Mode 2 (Resonant mode): (t_1-t_2)

After T_1 is turned off under ZVS condition, resonance between L_r and C_r occurs. $V_{cr}(t)$ decreases from V_s to 0. At t_2, $i_{Lr}(t)$ reaches the peak value in this interval. The equations are:

$$i_{Lr}(t) = \frac{V_s}{Z_r}\sin(\omega_r t) + [I_1 + (I_o - I_s)]\cos(\omega_r t) - (I_o - I_s)$$

$$V_{Cr}(t) = -V_s\cos(\omega_r t) - [I_1 + (I_o - I_s)]Z_r\sin(\omega_r t)$$

$$i_{Lr}(t_2) = I_2 = I_{Lr,peak}$$

$$V_{Cr}(t_2) = 0$$

where

$$Z_r = \sqrt{\frac{L_r}{C_r}} \text{ and } \omega_r = \frac{1}{\sqrt{L_r C_r}}$$

(4) Mode 3 (Freewheeling mode): (t_2-t_3)

The resonant inductor current flows through two freewheeling paths (T_2-L_r-D_2 and T_3-D_1-L_r). This duration is the zero voltage period created for ZVS of the inverter, and should be longer than the minimum on and off times of the inverter's power switches.

$$i_{Lr}(t) = I_2$$

$$v_{Cr}(t) = 0$$

(5) Mode 4 (Resonant mode): (t_3-t_4)

This mode begins when T_2 and T_3 are switched off under ZVS. The second half of the resonance between L_r and C_r starts again. The capacitor voltage $V_{cr}(t)$ increases back from 0 to V_s and is clamped to V_s. The relevant equations in this mode are

$i_{Lr}(t) = [I_2 - (I_{on} - I_s)] \cos(\omega_r t) - (I_{on} - I_s)$

$V_{Cr}(t) = [I_2 - (I_{on} - I_s)] Z_r \sin(\omega_r t)$

$i_{Lr}(t_4) = I_3$

$V_{Cr}(t_4) = V_s$

where I_{on} is the load current after the switching state.

(6) Mode 5 (Discharging mode): $(t_4 - t_5)$

In this period, T_1 is switched on under ZV condition because $V_{cr}(t) = V_s$. The inductor current decreases linearly. This mode finishes when $i_{Lr}(t)$ becomes zero.

$$i_{Lr}(t) = -\frac{V_s}{L_r}t + I_3$$

$$v_{Cr}(t) = V_s$$

$$i_{Lr}(t_5) = 0$$

Quasi-Resonant Soft-switched Inverter [50]

Circuit Operation

Consider an inverter fed by a dc voltage source vs a front-stage interface circuit shown in Fig. 55, can be added between the dc voltage source and the inverter. The front-stage circuit consists of a quasi-resonant circuit in which the first half of the resonance cycle is set to occur to create the zero-voltage condition whenever inverter switching is needed. After inverter switching has been completed, the second half of the resonance cycle takes place so that the dc link voltage is set back to its normal level. To avoid excessive losses in the resonant circuit, a small capacitor C_{r1} is normally used to provide the dc link voltage whilst the large smoothing dc link capacitor C_1 is isolated from the resonant circuit just before the zero-voltage duration. This method avoids the requirement for pulling the dc voltage of the bulk capacitor to zero

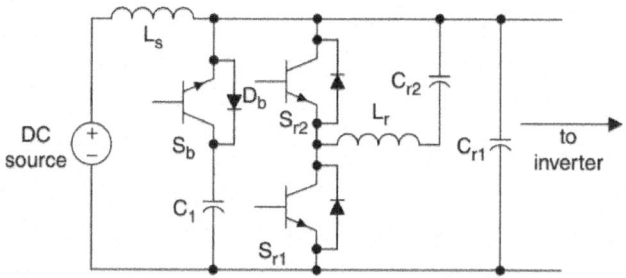

Figure 55: Quasi-resonant circuit for soft-switched inverter.

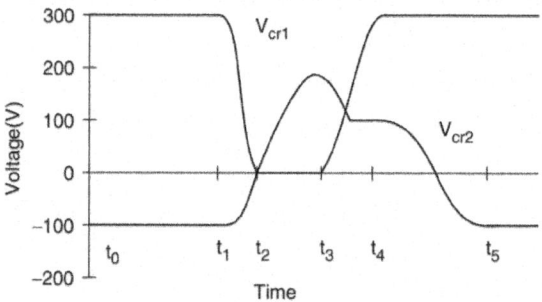

Figure 56: Typical waveforms for V_{cr1} and V_{cr2}.

The period for this mode is from t_0 to t_1 in Fig. 56. In this mode, switch Sb is turned on and switches S_{r1} and S_{r2} are turned off. The inverter in Fig. 55 works like a conventional dc link inverter. In this mode, $V_{cr1} = V_{c1}$. The voltage across switch Sb is zero. Before an inverter switching takes place, when switch S_{r1} is triggered at t_1 to discharge C_{r1}. This operating mode ends at t_2 when V_{cr1} approaches zero. The equivalent circuit in this mode is shown in Fig. 57a. The switch Sb must be turned off at zero voltage when switch S_{r1} is triggered. After S_{r1} is triggered, C_{r1} will be discharged via the loop C_{r1}, C_{r2}, L_r, and S_{r1}. Under conditions of $V_{cr2} \leq 0$ and $C_{r1} \leq C_{r2}$, the energy stored in C_{r1} will be transferred to C_{r2} and V_{cr1} falls to zero in the first half of the resonant cycle in the equivalent circuit of Fig. 57a. V_{cr1} will be clamped to zero by the freewheel diodes in the inverter bridge and will not become negative. Thus, V_{cr1} can be pulled down to zero for zero-voltage switching. When the current in inductor L_r becomes zero, switch S_{r1} can be turned off at zero current.

Inverter switching can take place in the period from t_2 to t_3 in which V_{cr1} remains zero. This period must be longer than the turn-on and turn-off times of the switches. When inverter switching has been completed, it is necessary to reset the voltage of capacitor C_{r1}. The equivalent circuit in this mode is shown in Fig. 57b. The current in inductor L_r reaches zero at t_3. Due to the voltage in V_{cr2} and the presence of diode D_r, this current then flows in the opposite direction.

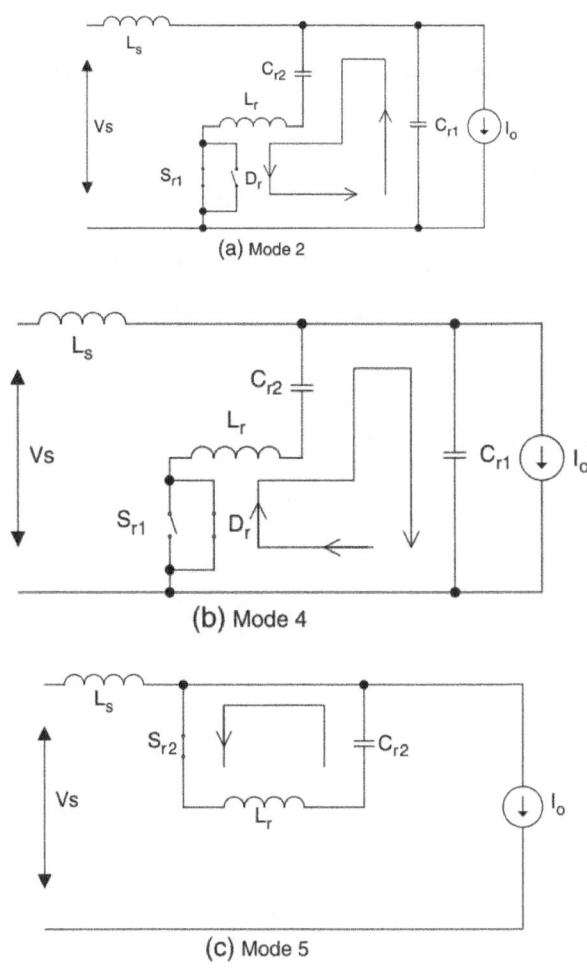

(a) Mode 2

(b) Mode 4

(c) Mode 5

Figure 57: Equivalent circuits of the quasi-resonant circuit for different modes.

C_{r1} will be recharged via L_r, C_{r2}, C_{r1}, and D_r. The diode D_r turns off when the current in L_r becomes zero. V_{cr1} will not go beyond 1 per unit because C_{r1} is clamped to supply voltage by diode D_b. The switch S_b can be turned on again at zero-voltage condition when V_{cr1} returns to normal dc supply voltage. After Dr turns off, V_{cr2} may not be zero. Some positive residual capacitor voltage remains in C_2 at t_4, as shown in Fig. 56. In case V_{cr2} is positive, V_{cr1} cannot be pulled down to zero again in the next switching cycle. Therefore, S_{r2} should be triggered after t_4 to reverse the residual voltage in C_{r2}. At time t_5, S_{r2} turns off at zero-current condition and V_{cr2} is now reversed to negative. The equivalent circuit in this mode is shown in Fig. 57c. When $V_{cr2} \leq 0$ and $C_{r1} \leq C_{r2}$, V_{cr1} can be pulled down to zero again before the next inverter switching. The operation can then be repeated in next switching cycle.

Design Considerations

(1) C_{r1} and C_{r2}

The criterion for getting zero capacitor voltages V_{cr1} is:

$$(C_{r1} - C_{r2})V_s + 2C_{r2}V_{o2} - \Delta I \pi \sqrt{L_r C_e} \leq 0 \tag{23}$$

Where

• V_{o1} is the initial voltage of C_{r1};

• V_{o2} is the initial voltage of C_{r2};

• i_{L0} is the initial current of inductor L_r;

• $I = I_o - I_s$, which is the difference between load current and supply current. It is assumed to be a constant within a resonant cycle;

• R_r is the equivalent resistance in the resonant circuit.

$$\bullet \quad C_e = \frac{C_{r1} C_{r2}}{C_{r1} + C_{r2}}$$

When $I \geq 0$, the above criterion is always true under conditions of:

$$C_{r1} \leq C_{r2}, \quad V_{o2} \leq 0$$

The criterion for recharging voltage V_{cr1} to 1 per unit dc link voltage is:

$$\frac{2C_{r2}}{C_{r1} + C_{r2}} V_{o2} - \frac{\Delta I}{C_{r1} + C_{r2}} \pi \sqrt{L_r C_e} \geq V_s \tag{24}$$

(2) Inductor L_r

The inductor L_r should be small so that the dc link voltage can be decreased to zero quickly. However, a small L_r could result in large peak resonant current and therefore requirement of power devices with large current pulse ratings. An increase in the inductance of L_r can limit the peak current in the quasiresonant circuit. Because the resonant frequency depends on both the inductor and the capacitor, therefore, the selection of L_r can be considered together with the capacitors C_{r1} and C_{r2} and with other factors such as the current ratings of power devices, the zero-voltage duration and the switching frequency required in the soft-switching circuit.

(3) Triggering instants of the switches

The correct triggering instants for the switches are essential for the successful operation of this soft-switched inverter. For the inverter switches, the triggering instants are determined from a PWM modulation. Let T_s be the time at which the inverter switches change states. To get the zero-voltage inverter switching, switch S_{r1} should be turned on half resonant cycle before the inverter switching instant. The turn-on instant of S_{r1}, which is t_1 in Fig. 56, can be written as:

$$t_1 = T_s - \frac{\pi}{\omega} \qquad (25)$$

where $\omega = \sqrt{\omega_0^2 - \alpha^2}$, $\alpha = \frac{R_r}{2L_r}$, $\omega_0 = \sqrt{\frac{1}{L_r C_e}}$. The switch S_b is turned off at t_1.

S_{r1} may be turned off during its zero current period when diode D_r is conducting. For easy implementation, its turn-off time can be selected as $T_s + \pi/\omega$. Because the d_c link voltage can be pulled down to zero in less than half resonant cycle, T_s should occur between t_2 and t_3.

At time t_3 (the exact instant depends on the I), the diode D_r turns on in the second half of the resonant cycle to recharge C_{r1}. At t_4, V_{cr1} reaches 1 per unit and diode D_b clamps V_{cr1} to 1 per unit. The switch S_b can be turned on again at t_4, which is half resonant cycle after the start of t_3:

$$t_4 \approx t_3 + \frac{\pi}{\omega} \qquad (26)$$

As t_3 cannot be determined accurately, a voltage sensor in principle can be used to provide information for t_4 so that S_b can be turned on to reconnect C_1 to the inverter. In practice, however, S_b can be turned on a few microseconds (longer than $T_s + \pi/\omega$) after t_2 without using a voltage sensor (because it is not critical for S_b to be on exactly at the moment V_{cr} reaches the nominal voltage). As for switch S_{r2}, it can be turned on a few microseconds after t_4. It will be turned off half of resonant cycle ($\pi \sqrt{L r C_{r2}}$) in the L_r–C_{r2} circuit later. In practice, the timing of S_{r1}, S_{r2}, and Sb can be adjusted in a simple tuning procedure for a given set of parameters. Figure 58 shows the measured gating signals of S_{r1}, S_{r2}, and S_b with the d_c link voltage V_{cr1} in a 20 kHz switching inverter. Figures 59 and 60 show the measured waveforms of I_s, I_o, and V_{crl} under no load condition and loaded condition, respectively.

Control of Quasi-resonant Soft-switched Inverter Using Digital Time Control (DTC) [51]

Based on the zero-average-current error (ZACE) control concept, a digital time control method has been developed for a current-controlled quasi-resonant soft-switched inverter. The basic ZACE concept is shown in Fig. 61. The current error is obtained from the difference of a reference current and the sensed current. The idea is to make the areas of each transition (A1 and A2) equal. If the switching frequency is significantly greater than the fundamental frequency of the reference signal,

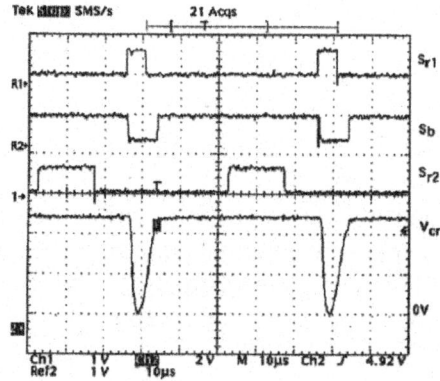

Figure 58: Gating signals for S_{r1}, S_{r2}, and S_b with V_{crl}.

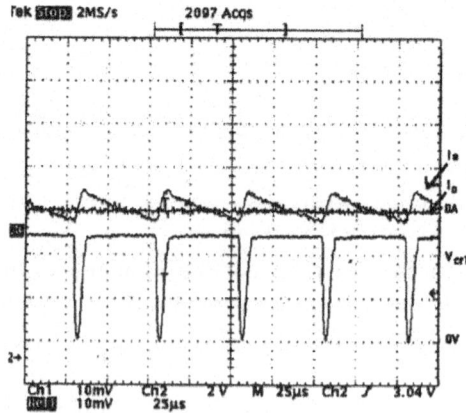

Figure 59: Typical I_s, I_o, and V_{crl} under no-load condition.

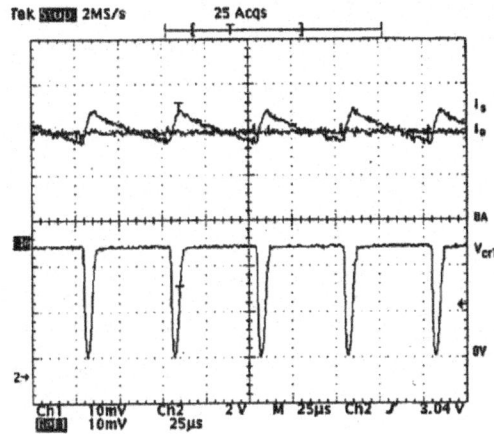

Figure 60: Typical I_s, I_o, and V_{crl} under loaded condition.

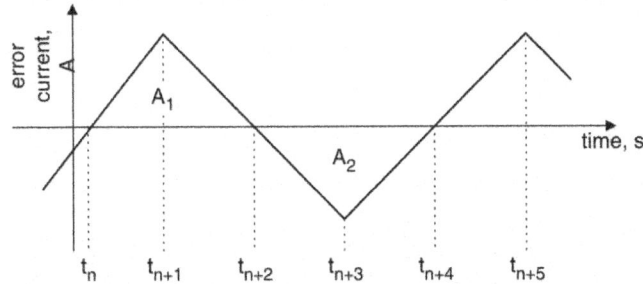

Figure 61: Zero-average-current error (ZACE) control concept.

the rising and falling current segments can be assumed to be linear. The following simplified equation can be established.

$$\Delta t_{n+1} = t_{n+1} - t_n$$

(27)

The control algorithm for the inverter is

$$\Delta t_{n+3} = \Delta t_{n+2} + D\left[\frac{T_{sw}}{2} - (t_{n+2} - t_n)\right]$$

(28)

where $D = \dfrac{\Delta t_{n+2}}{t_{n+2} - t_n}$ and $T_{sw} = t_{n+4} - t_n$.

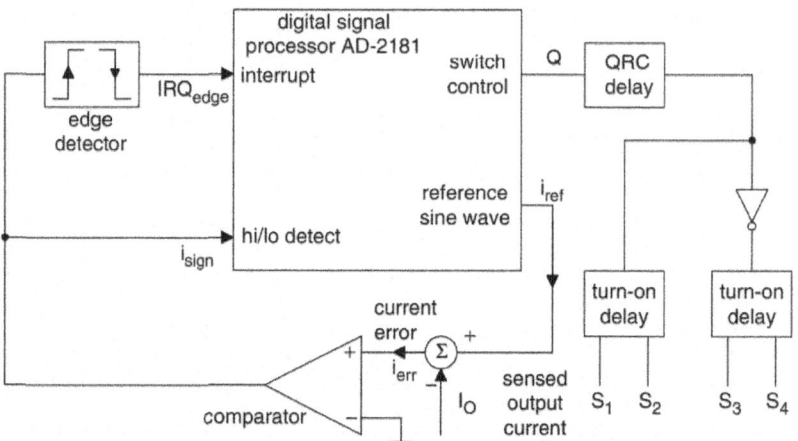

Figure 62: Implementation of DTC.

The schematic of a digital signal processor (DSP) based controller for the DTC method is shown in Fig. 62. The duty cycle can be approximated from the reference sine wave by level shifting and scaling it between 0 and 1. The time $t_{n+2} - t_n$ is the sum of t_{n+1} and t_{n+2}. These data provide information for the calculation of the next switching time t_{n+1}.

The switches are triggered by the changing edge of the switch control Q. Approximate delays are added to the individual switching signals for both the inverter switches and the quasi-resonant switches. Typical gating waveforms are shown in Fig. 63. The use of the quasi-resonant soft-switched inverter is a very effective way in suppressing switching transient and EMI emission. Figures 64a,b show the inverter switch voltage waveforms of a standard hard-switched inverter and a quasi-resonant soft-switched inverter, respectively. It is clear that the soft-switched waveform has much less transient than the hard-switched waveform.

Resonant Pole Inverter (RPI) and Auxiliary Resonant Commutated Pole Inverter (ARCPI)

The resonant pole inverter integrates the resonant components with the output filter components L_f and C_f. The load is connected to the mid-point of the dc bus capacitors as shown in Fig. 65.

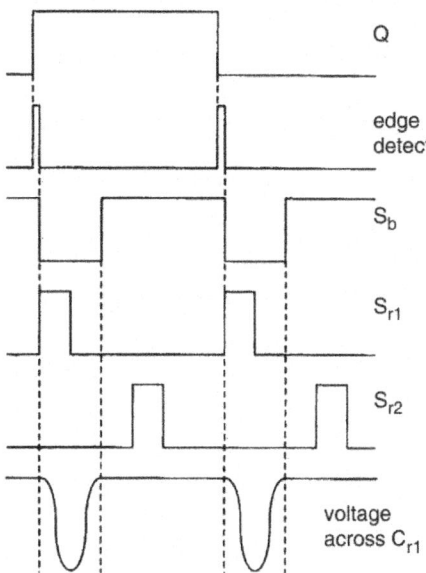

Figure 63: Timing diagrams for the gating signals.

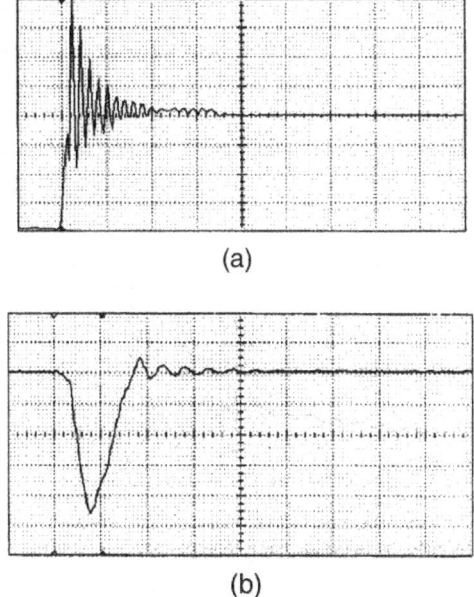

Figure 64: (a) Typical switch voltage under hard turn-off and (b) typical switch voltage under soft turn-off.

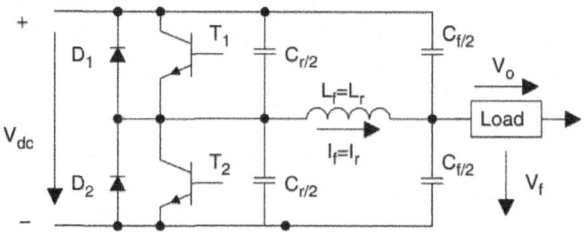

Figure 65: One leg of a resonant pole inverter.

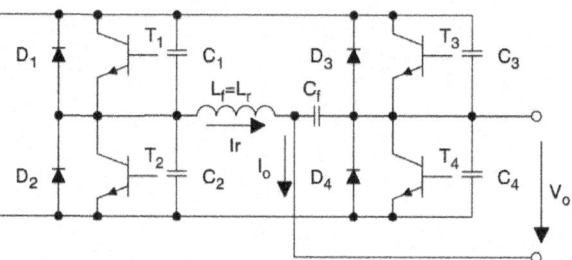

Figure 66: Single-phase resonant pole inverter.

It should however be noted that the RPI can be described as a resonant inverter. Figure 66 shows a single-phase RPI. Its operation can be described with the timing diagram in Fig. 67. The operating modes are included in Fig. 68. The RPI provides soft-switching for all power switches. But it has two disadvantages. First, the power devices have to be switched continuously at the resonant frequency determined by the resonant components. Second, the power devices in the RPI circuit require a 2.2–2.5 p.u. current turn-off capability.

An improved version of the RPI is the auxiliary resonant commutated pole inverter (ARCPI). The ARCPI for one inverter leg is shown in Fig. 69. Unlike the basic RPI, the ARCPI allows the switching frequency to be controlled. Each of the primary switches is closely paralleled with a snubber capacitor to ensure ZV turn off. Auxiliary switches are connected in series with an inductor, ensuring that they operate under ZC conditions. For each leg, an auxiliary circuit comprising two extra switches A1 and A2, two freewheeling diodes, and a resonant inductor L_r is required. This doubles the number of power switches when compared with hard-switched inverters. Figure 70 shows the three-phase ARCPI system. Depending on the load conditions, three commutation modes are generally needed. The commutation methods at low and high current are different. This makes the control of the ARCPI very complex. The increase in control and circuit complexity represents a considerable cost penalty [52, 53].

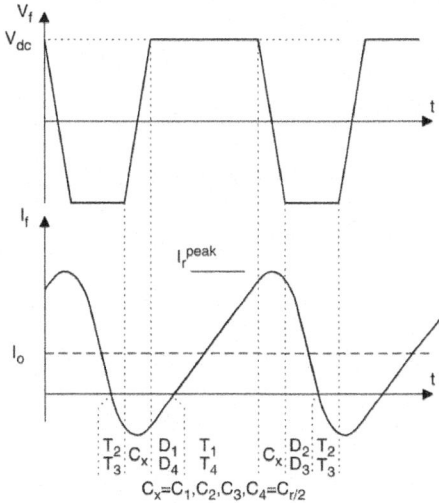

Figure 67: Timing diagram for a single-phase resonant pole inverter.

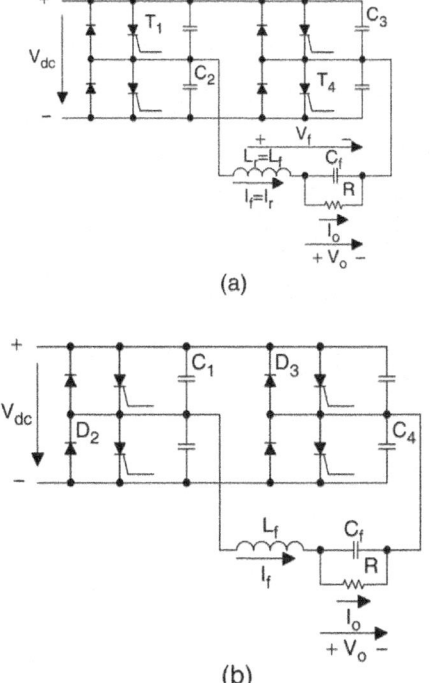

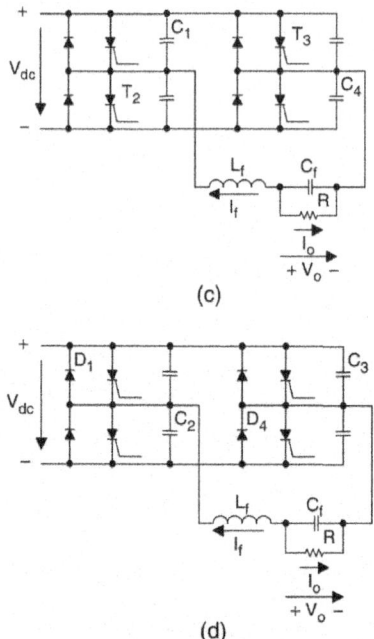

(c)

(d)

Figure 68: Operating modes of a single-phase resonant pole inverter.

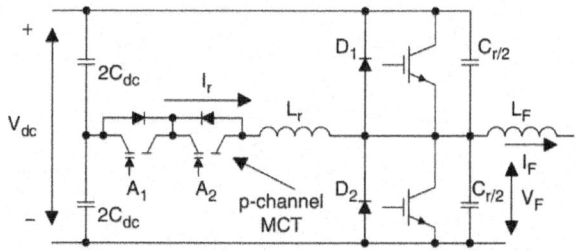

Figure 69: Improved resonant pole inverter leg.

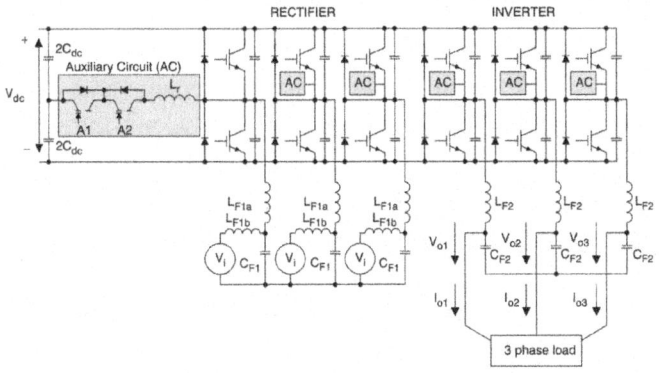

Figure 70: Three-phase auxiliary resonant commutated pole inverter (ARCPI).

REFERENCES

1. R. M. Davis, Power Diode and Thyristor Circuits, IEE Monograph, Series 7, Herts: Peregrinus, 1971.

2. N. Mohan, T. Undeland, and W. Robbins, Power Electronics, Converters, Applications, and Design, Hoboken, NJ: John Wiley and Sons, 1995.

3. M. H. Rashid, Power Electronics, Circuits, Devices, and Applications, Upper Saddle River, NJ: Pearson/Prentice Hall, 2004.

4. P. Vinciarelli, "Forward Converter Switching at Zero Current," U.S. Patent 4,416,959, November 1983.

5. F. C. Lee, High-Frequency Resonant, Quasi-Resonant and MultiResonant Converters, Blacksburg, VA: Virginia Power Electronics Center, 1991.

6. F. C. Lee, High-Frequency Resonant and Soft-Switching Converters, Virginia Power Electronics Center, 1991.

7. K. Kit Sum, Recent Developments in Resonant Converters, Ventura, CA, Intertec Communication Press, 1988.

8. R. E. Tarter, Solid-State Power Conversion Handbook, New York WileyInterscience publication, 1993.

9. K. H. Liu and F. C. Lee, "Resonant Switches – A Unified Approach to Improve Performances of Switching Converters," in Proc. Int. Telecomm. Energy Conf., 1984, pp. 344–351.

10. K. H. Liu, R. Oruganti, and F. C. Lee, "Resonant Switches – Topologies and Characteristics," in Proc. IEEE Power Electron. Spec. Conf., 1985, pp. 62–67.

11. K. D. T. Ngo, "Generalized of Resonant Switches and Quasi-Resonant

DC-DC Converters," in Proc. IEEE Power Electron. Spec. Conf., 1986, pp. 58–70.

12. F. C. Lee, "High-frequency Quasi-Resonant and Multi-Resonant Converter Technologies," in Proc. IEEE Int. Conf. Ind. Electron., 1988, pp. 509–521.

13. K. H. Liu and F. C. Lee, "Zero-Voltage Switching Techniques in DC/DC Converter Circuits," in Proc. IEEE Power Electron. Spec. Conf., 1986, pp. 58–70.

14. M. Jovanovic, W. Tabisz, and F. C. Lee, "Zero-voltage Switching Technique in High Frequency Off-line Converters," in Proc. Applied Power Electron. Conf. and Expo., 1988, pp. 23–32.

15. R. Steigerwald, "A Comparison of Half-Bridge Resonant Converter Topologies," IEEE Trans. Power Electron., vol. 3, no. 2, April. 1988, pp. 174–182.

16. O. D. Patterson and D.M. Divan, "Pseudo-resonant Full-Bridge DC/DC Converter," in Proc. IEEE Power Electron. Spec. Conf., 1987, pp. 424–430.

17. C. P. Henze, H. C. Martin, and D. W. Parsley, "Zero-Voltage Switching in High Frequency Power Converters Using Pulse Width Modulation," in Proc. IEEE Applied Power Electron. Conf. and Expo., 1988, pp. 33–40.

18. General Electric Company, "Full-Bridge Lossless Switching Converters," U.S. Patent 4,864,479, 1989.

19. W. A. Tabisz and F. C. Lee, "DC Analysis and Design of Zero-Voltage Switched Multi-Resonant Converters," in Proc. IEEE Power Electron. Spec. Conf., 1989, pp. 243–251.

20. W. A. Tabisz and F. C. Lee, "Zero-Voltage-Switching Multi-Resonant Technique – a Novel Approach to Improve Performance of High Frequency Quasi-Resonant Converters," IEEE Trans. Power Electron., vol. 4, no. 4, October 1989, pp. 450–458.

21. M. Jovanovic and F. C. Lee, "DC Analysis of Half-Bridge ZeroVoltage-Switched Multi-Resonant Converter," IEEE Trans. Power Electron., vol. 5, no. 2, April 1990, pp. 160–171.

22. R. Farrington, M. M. Jovanovic, and F. C. Lee, "Constant-Frequency Zero-Voltage-Switched Multi-resonant converters: Analysis, Design, and Experimental Results," in Proc. IEEE Power Electron. Spec. Conf., 1990, pp. 197–205.

23. G. Hua, C. S. Leu, and F. C. Lee, "Novel zero-voltage-transition PWM converters," in Proc. IEEE Power Electron. Spec. Conf., 1992, pp. 55–61.

24. R. Watson, G. Hua, and F. C. Lee, "Characterization of an Active Clamp Flyback Topology for Power Factor Correction Applications," IEEE Trans. Power Electron., vol. 11, no. 1, January 1996, pp. 191–198.

25. R. Watson, F. C. Lee, and G. Hua, "Utilization of an ActiveClamp Circuit to Achieve Soft Switching in Flyback Converters," IEEE Trans. Power Electron., vol. 11, no. 1, January 1996, pp. 162–169.

26. B. Carsten, "Design Techniques for Transformer Active Reset Circuits at High Frequency and Power Levels," in Proc. High Freq. Power Conversion Conf., 1990, pp. 235–245.

27. S. D. Johnson, A. F. Witulski, and E. W. Erickson, "A Comparison of Resonant Technologies in High Voltage DC Applications," in Proc. IEEE Appl. Power Electron. Conf., 1987, pp. 145–166.

28. A. K. S. Bhat and S. B. Dewan, "A Generalized Approach for the Steady State Analysis of Resonant Inverters," IEEE Trans. Ind. Appl., vol. 25, no. 2, March 1989, pp. 326–338.

29. A. K. S. Bhat, "A Resonant Converter Suitable for 650V DC Bus Operation," IEEE Trans. Power Eletcron., vol. 6, no. 4, October 1991, pp. 739–748.

30. A. K. S. Bhat and M. M. Swamy, "Loss Calculations in Transistorized Parallel Converters Operating Above Resonance," IEEE Trans. Power Electron., vol. 4, no. 4, July 1989, pp. 449–458.

31. A. K. S. Bhat, "A Unified Approach for the Steady-State Analysis of Resonant Converter," IEEE Trans. Ind. Electron., vol. 38, no. 4, August 1991, pp. 251–259.

32. Applications Handbook, Unitrode Corporation, 1999.

33. Barbi, J. C. Bolacell, D. C. Martins, and F.B. Libano, "Buck Quasiresonant Converter Operating at Constant Frequency: Analysis, Design and Experimentation," PESC'89, pp. 873–880.

34. K. W. E. Cheng and P. D. Evans, "A Family of Extended-period Circuits for Power Supply Applications using High Conversion Frequencies," EPE'91, pp. 4.225–4.230.

35. S. Y. R. Hui, K. W. E. Cheng, and S. R. N. Prakash, "A Fully Softswitched Extended-period Quasi-resonant Power Correction Circuit," IEEE Transactions on Power Electronics, vol. 12, no. 5, September 1997, pp. 922–930.

36. H. S. H. Chung, S. Y. R. Hui, and W. H. Wang, "A Zero-CurrentSwitching PWM Flyback Converter with a Simple Auxiliary Switch," IEEE Transactions on Power Electronics, vol. 14, no. 2, March 1999, pp. 329–

342.

37.   H. S. H. Chung and S. Y. R. Hui, "Reduction of EMI in Power Converters Using fully Soft-switching Technique," IEEE Transactions on Electromagnetics, vol. 40, no. 3, August 1998, pp. 282–287.

38.   H. S. H. Chung, W. L. Cheung, and K. S. Tang, "A ZCS Bidirectional Flyback Converter," IEEE Transactions on Power Electronics, vol. 19, no. 6, November 2004, pp. 1426–1434.

39.   M. T. Zhang, M. M. Jovanovic, and F. C. Lee, "Design Considerations and Performance Evaluation of Synchronous Rectification in Flyback Converter," IEEE Transactions on Power Electronics, vol. 13, no. 3, May 1998, pp. 538–546.

40.   T. Undeland, "Snubbers for Pulse Width Modulated Bridge Converters with Power Transistors or GTOs," IPEC, Tokyo, Conference proceedings, vol. 1, 1983, pp. 313–323.

41.   McMurray, "Efficient snubbers for voltage source GTO inverters," IEEE Transactions on Power Electronics, vol. 2, no. 3, July 1987, pp. 264–272.

42.   D. M. Divan, "The resonant dc link converter-A new concept in static power conversion," IEEE Trans. IA, vol. 25, no. 2, 1989, pp. 317–325.

43.   Jin-Sheng Lai and B. K. Bose, "An Induction Motor Drive Using an Improved High Frequency Resonant DC Link Inverter," IEEE Trans. on Power Electronics, vol. 6, no. 3, 1991, pp. 504–513.

44.   D. M. Divan and G. Skibinski, "Zero-Switching-Loss Inverters for High-Power Applications," IEEE Trans. IA, vol. 25, no. 4, 1989, pp. 634–643.

45.   S. J. Finney, T. C. Green, and B. W. Williams, "Review of Resonant Link Topologies for Inverters," IEE Proc. B, vol. 140, no. 2, 1993, pp. 103–114.

46.   S. B. Dewan and D. L. Duff, "Optimum Design of an Input Commutated Inverter for AC Motor Control," IEEE Trans. on Industry Gen. Applications, vol. IGA-5, no. 6, November/December 1969, pp. 699–705.

47.   V.R. Stefanov and P. Bhagwat, "A Novel DC Side Commutated Inverter," IEEE PESC Record, 1980.

48.   A. Kurnia, H. Cherradi, and D. Divan, "Impact of IGBT behavior on design optimization of soft switching inverter topologies," IEEE IAS Conf. Record, 1993, pp. 140–146.

49.   Y. C. Jung, J. G. Cho, and G. H. Cho, "A New Zero Voltage Switching Resonant DC-Link Inverter with Low Voltage Stress," Proc. of IEEE Industrial Electronics Conference, 1991, pp. 308–13.

50. S. Y. R. Hui, E. S. Gogani and J. Zhang, "Analysis of a Quasiresonant Circuit for Soft-switched Inverters," IEEE Transactions on Power Electronics, vol. 11, no. 1, January, 1999, pp. 106–114.

51. D. M. Baker, V. G. Ageliidis, C. W. Meng, and C. V. Nayar, "Integrating the Digital Time Control Algorithm with DC-Bus 'Notching' Circuit Based Soft-Switching Inverter," IEE Proceedings-Electric Power Applications, vol. 146, no. 5, September 1999, pp. 524–529.

52. F. C. Lee and D. Borojevic, "Soft-switching PWM Converters and Inverters," Tutorial notes, PESC'94.

53. D. M. Divan and R. W. De Doncker, "Hard and Soft-switching Voltage Source Inverters", Tutorial notes, PESC'94.

CITATION

CHAPTER 1

M. S. Ali, S. K. Kamarudin, M. S. Masdar, and A. Mohamed, "An Overview of Power Electronics Applications in Fuel Cell Systems: DC and AC Converters," The Scientific World Journal, vol. 2014, Article ID 103709, 9 pages, 2014. doi:10.1155/2014/103709.

CHAPTER 2

Ioannis Ch. Proimadis, Dionysios V. Spyropoulos, and Epaminondas D. Mitronikas, "An Alternative for All-Electric Ships Applications: The Synchronous Reluctance Motor," Advances in Power Electronics, vol. 2013, Article ID 862734, 7 pages, 2013. doi:10.1155/2013/862734.

CHAPTER 3

A. Zebda, S. Cosnier, J.-P. Alcaraz, M. Holzinger, A. Le Goff, C. Gondran, F. Boucher, F. Giroud, K. Gorgy, H. Lamraoui & P. Cinquin, Single Glucose Biofuel Cells Implanted in Rats Power Electronic Devices, doi:10.1038/srep01516.

CHAPTER 4

C. Carretero, J. Acero, and R. Alonso, TM-TE Decomposition of Power Losses in Multi-Stranded LITZ-Wires used in Electronic Devices, Vol. 123, 83–103, 2012, http://www.jpier.org/PIER/pier123/06.11091909.pdf.

INDEX